Engineering Essentials

Engineering Essentials

Freddie Davis

Kenneth Leitch

cognella® | ACADEMIC PUBLISHING

The Engineering Profession

Objectives

- Define the term *engineering*.
- Identify the primary engineering disciplines.
- Identify and describe topics that engineers/engineering technologists must study for an ABET-accredited degree.
- Identify ABET Criterion 3 and their application to engineering courses.
- Describe career and continuing education requirements.

Engineering Defined

One long definition of *engineering* is that it is the discipline, art, and profession of utilizing and applying technical, scientific, and mathematical knowledge to design and implement materials, structures, machines, devices, systems, and processes (hereafter called *engineered systems* or *products*) that meet required specifications in order to safely realize a desired objective. A shorter variant of this definition is that engineers use mathematics and science in order to create something that is useful.

According to *ABET* (formerly known as *The Accreditation Board for Engineering and Technology*, www.abet.org), there are more than 770 schools that offer over 3,800 engineering and engineering technology degree programs in the United States and 32 other countries. Engineering degrees require a background in math, science, technical writing and communication, and specialized engineering coursework. Engineering is considered a *professional field*, which is a vocation or career founded on specialized educational training in order to provide services and goods for a direct and definite compensation (salary or pay) wholly apart from expectation of other business gain. Other professional fields include law, business, medicine, nursing, and clergy. Professional fields frequently require accredited education and licensing by government to protect the public, a topic that will be explored later.

Engineers work together with many other professionals. Scientists examine and test phenomena, creating new knowledge, and engineers in multiple subdisciplines (which will be described shortly) employ scientific knowledge and mathematics to utilize materials and the forces of nature to benefit society. Engineering technologists apply the work of scientists and engineers to get tangible results; they often work in the field to implement technological applications. Engineers and engineering technologists must write clear specifications for skilled workers to follow to construct machinery, vehicles, electrical equipment, buildings, water systems, roads, and many other designs.

Engineers and engineering technologists have a wide variety of tasks that they will perform in their careers. Many are involved in the research, design, and development of new engineered systems. Others will create prototypes of engineered systems and test them to determine whether they meet required specifications. An engineer may be responsible for operation and management of a firm or work to sell the firm's product to government and/or consumers. Engineers are highly skilled persons that can also become consultants or teachers, especially after several years of practice in the engineering field.

Engineering has over 20 distinct disciplines. According to the American Society for Engineering Education (ASEE), the top five disciplines in order by graduates during 2015/2016 were mechanical, computer science (within engineering), electrical, civil, and chemical engineering. In total, approximately 112,000 students graduated with a bachelor's degree, 62,000 graduated with a master's degree, and over 11,000 graduated with a doctoral degree across all engineering fields during 2015/2016 in the USA.

Overview of Popular Engineering Majors

Mechanical engineering is a very broad discipline. Several other disciplines are often linked to mechanical, including that of aerospace, nuclear, and ocean, to name a few. As evidenced by its name, mechanical engineering has a major emphasis on movable machines, including vehicles, aircraft, and manufacturing. Mechanical (and its offshoot, aerospace) engineering arose during the Industrial Revolution of the eighteenth century, which is when the discipline became distinct from civil engineering. Mechanical engineering has a large number of subject areas, including thermal sciences, solid mechanics, manufacturing, mechatronics, materials science, and measurements. A student studying mechanical engineering can expect a wide variety of career opportunities, including energy generation, machine design and development, control of automated systems, manufacturing and processing of materials, and solutions to environmental problems. Many mechanical (as well as aerospace) engineers work in the aerospace and defense industry. Such engineers design aircraft, missiles, space vehicles, satellites, ships, and submarines. The US Bureau of Labor Statistics (http://www.bls.gov/) forecasts a 9% increase in growth between 2016 and 2026 for this discipline, with approximately 288,800 in the workforce in the United States in 2016.

Civil engineering is the oldest and broadest branch of engineering, with the term having been devised to distinguish its practitioners from *military* engineers. The first civil (alongside military) engineering program in the United States was established at the West Point Military Academy in 1802 and continues to the present. Civil engineering involves the application of math and science to design, construct, and operate the infrastructure necessary for modern life: buildings, bridges, roads and transport, water treatment, power generation and distribution, and much more. Six major branches of civil engineering are generally recognized and are normally part of university course studies. *Structural engineering* is the design and construction of buildings, bridges, dams, tunnels, and support structures. *Transportation engineering* is the design and operation of facilities for land, water, and air. *Environmental engineering* is the design and operation of facilities for drinking water and wastewater treatment, as well as the control of pollution in land, water, and air. *Water resources engineering* involves flood control and hydrological studies. *Geotechnical engineering* involves foundation design, earthquake design, soil analysis, groundwater analysis, and pavement design. *Surveying and construction engineering* involves the development of topographical maps, legal descriptions of land, and routing of transportation facilities, as well as estimating costs, supervision, start-up, and initial operation of civil engineering facilities. The US Bureau of Labor Statistics (http://www.bls.gov/) forecasts an 11% increase in growth between 2016 and 2026 for this discipline, with approximately 303,500 in the workforce in the United States in 2016.

Electrical engineers are responsible for a wide variety of modern technologies that have emerged in the last one hundred years. In particular, these engineers are responsible for communication technologies such as telephone, radio, radar, computer, internet, cellular telephone networks (a type of radio communication), and others that make instantaneous exchange of information possible. Many of these technologies require both hardware and software solutions to work effectively. These engineers also make the power grid and batteries that supply electricity to these devices possible. Electrical and computer engineers deal with both analog (physical media representation) and digital (numerical representation) of data. Examples of analog versus digital data representation include for sound (a vinyl record versus a CD or MP3 file) and text (a book versus an e-reader or computer file). The US Bureau of Labor Statistics reports approximately 324,600 electrical engineers in the workforce in the United States in 2016, with growth estimated at 7% between 2016 and 2026.

Chemical engineers play an important role in creating the materials that other engineers use for their own projects. These engineers design, create, and operate processes to produce construction materials, manufacturing materials, fuel, food products, plastics, fertilizers, healthcare products, fibers (as for cloth or construction materials), and many others. These engineers must have knowledge of chemical processes (organic, inorganic, biochemical, etc.), safety, material and energy balance, thermodynamics, heat transfer, chemical reactions, separation techniques, process design, and experimental techniques. The US Bureau of Labor Statistics reports approximately 32,700 chemical engineers in the workforce in the United States in 2016.

Environmental engineering began as an integral subarea of civil engineering and has overlap with chemical, mechanical, and agricultural engineering. Environmental engineers help to clean up air, water, and land pollution, as well as mitigate noise effects. Agricultural production of crops and livestock, as well as manufacturing and energy production, produces large amounts of contamination that must be mitigated before waste is discharged into the environment. Environmental engineers design the processes for these management systems. They also help to deal with natural disasters, such as flooding and earthquakes. These engineers are devising ways to use or reuse waste materials through recycling. For example, recycling of metals and plastics can be beneficial to the environment, as well as providing materials for new products. Other waste materials, such as fly ash from the burning of coal, are now used for concrete production instead of being sent to a landfill, a benefit to society as the result of environmental engineering research. The US Bureau of Labor Statistics forecasts an expected 8% increase in growth between 2016 and 2026 for this discipline and reports approximately 53,800 environmental engineers in the workforce in the United States in 2016.

Biomedical engineering began as an offshoot of the mechanical and chemical engineering disciplines. Biomedical engineers work with medical professionals to create surgical implants and prosthetic devices. They also study how the human body carries load effects and how to keep the body safe in the event of crash events, such as those that occur in automobiles, airplanes, and other modes of transportation. The US Bureau of Labor Statistics reports 21,300 biomedical engineers in the workforce in the United States in 2016.

Industrial engineers work to improve manufacturing processes, as well as the ergonomics (i.e., use) of engineered systems. These engineers improve the competitiveness of businesses, governments, and nonprofit entities by applying theory to human endeavors (i.e., helping us to work better). Industrial engineers work in manufacturing to oversee operations, product specifications, product conformity to specifications, layouts, and implementation. These engineers also study human factors to ensure human comfort with computer interfaces, machinery, furniture, vehicles, and other engineering systems designed for human use. Industrial engineers help to make financial decisions and conduct probability and statistical studies, quality control, experimentation, mathematical modeling, simulations, and programming, as well as overseeing the development of applications. The US Bureau of Labor Statistics reports approximately 25,100 industrial engineers in the workforce in the United States in 2016.

Comparison of Engineering and Engineering Technology

Engineering technologists are specialists that implement or extend existing technology, often of a specific engineering discipline. They often play a supporting role to engineers using basic engineering principles and technical skills. Engineering technologists (often called ETs) frequently work to improve products, manufacturing, construction, and operational functions. Students often wonder what the similarities and differences between engineering technology (ET) and other engineering fields are. The *National Society of Professional Engineers (NSPE)* describes the differences in this manner (from www.nspe.org):

> The distinction between engineering and engineering technology emanates primarily from differences in their educational programs. Engineering programs are geared toward development of conceptual skills, and consist of a sequence of engineering fundamentals and design courses, built on a foundation of complex mathematics and science courses. Engineering technology programs are oriented toward application, and provide their students introductory mathematics and science courses, and only a qualitative introduction to engineering fundamentals. Thus, engineering programs provide their graduates a breadth and depth of knowledge that allows them to function as designers. Engineering technology programs prepare their graduates to apply others' designs.

ABET certifies the following engineering and engineering technology programs and applied science programs through its commissions:

- Applied Science Accreditation Commission (ASAC), which accredits an applied science program at the associate's, bachelor's, or master's degree level.
- Computing Accreditation Commission (CAC), which accredits a computing program at the bachelor's degree level only.
- Engineering Accreditation Commission (EAC), which accredits an engineering program at the bachelor's or master's degree level.
- Engineering Technology Accreditation Commission (ETAC), which accredits an engineering technology program at the associate's or bachelor's degree level.

According to ABET (www.abet.org), the main differences between engineering and engineering technology programs are the following:

- Engineering undergraduate programs include more mathematics work and higher level mathematics than technology programs.
- Engineering undergraduate programs often focus on *theory*, while technology programs usually focus on *application*.
- Once they enter the workforce, engineering graduates typically spend their time planning, while engineering technology graduates spend their time making plans work.
- At ABET, engineering and engineering technology programs are evaluated and accredited by two separate accreditation commissions using two separate sets of accreditation criteria.
- Graduates from engineering programs are called *engineers*, while graduates of engineering technology programs are often called *engineering technologists*.
- Some U.S. state boards of professional engineering licensure will allow only graduates of engineering programs—not engineering technology programs—to become licensed engineers.

ABET Criterion 3

ABET criteria are important for engineering programs. In particular, the ABET criteria are divided into eight parts. The third criterion (changed by ABET from the long standing a through k prior to the 2019-2020 accreditation cycle) is addressed specifically in engineering programs and is reflected in the courses offered. The seven parts of Criterion 3 are as follows (www.abet.org):

1. an ability to identify, formulate, and solve complex engineering problems by applying principles of engineering, science, and mathematics
2. an ability to apply engineering design to produce solutions that meet specified needs with consideration of public health, safety, and welfare, as well as global, cultural, social, environmental, and economic factors
3. an ability to communicate effectively with a range of audiences

4. an ability to recognize ethical and professional responsibilities in engineering situations and make informed judgments, which must consider the impact of engineering solutions in global, economic, environmental, and societal contexts
5. an ability to function effectively on a team whose members together provide leadership, create a collaborative and inclusive environment, establish goals, plan tasks, and meet objectives
6. an ability to develop and conduct appropriate experimentation, analyze and interpret data, and use engineering judgment to draw conclusions
7. an ability to acquire and apply new knowledge as needed, using appropriate learning strategies

Engineers in Professional Practice

An engineering degree in accordance with ABET and university requirements requires a broad range of topics. The university has a set core of general education requirements in social sciences, economics, history, English, speech, political science, and the arts. Engineering and engineering technology degrees require physical sciences (chemistry, physics, biology, and/or geology), mathematics (algebra, trigonometry, calculus, differential equations, linear algebra, statistics, and/or numerical analysis), engineering core (fundamentals, graphics, computer programming, statics, mechanics, material science, ethics, and/or engineering economics), and engineering specialty courses that are degree-dependent.

When an engineer or engineering technologist graduates, his or her knowledge base can quickly become dated due to new technologies and methodologies constantly being developed. Continuing education is critical after obtaining a bachelor's degree. Many employers will pay for their engineers to attend conferences and seminars that provide the opportunity to obtain continuing education units (CEUs), which are required for licensed professional engineers. Advanced coursework and degrees (master's, PhD, MBA, etc.) also are helpful, especially when engineers wish to advance in their careers and to move into administration roles.

Engineers are professionals. As such, engineers have specialized knowledge and skills that can be used to benefit humanity. Engineers are held in regard by society due to the trust society has in their creations, many of which are essential for human health and life. Engineers should strive to be impartial in their service and to continually improve themselves and their chosen career field. Engineers should support their professional and technical societies, which in turn will represent them in matters of public and governmental importance.

Engineers may be required to obtain professional registration, especially when their engineered systems will be used in a public setting. To be registered, there are several conditions that must be met. First, in most states, one has to graduate in good standing from an ABET-accredited institution. Next, the engineer must successfully complete the computer-based *Fundamentals of Engineering (FE) exam* (www.ncees.org), becoming an *Engineer-in-Training (EIT)*. Then, the engineer must work a few years, typically four, as an EIT, with advanced degrees counting one year apiece toward this requirement. Finally, the engineer must successfully complete the *Principles and Practice exam*, becoming a *Professional Engineer (PE)*. States often also require PEs to pass an ethics exam and demonstrate continuous learning related to engineering.

Engineers may join professional societies, which advocate on their behalf. The discipline-specific professional societies also produce technical literature and outline professional standards. In particular, the ASCE, ASME, SAE, and IEEE (see "Discipline-specific" below) produce standards that are used in the creation of engineered systems and disseminate knowledge through journals, periodicals, and conferences. Other professional societies, such as SWE, SHPE, and NSBE (see "General interest" below), promote careers in groups that have been traditionally underrepresented in the engineering field. Below are the lists of engineering societies and their websites; students are encouraged to learn about and to join professional societies to network, find work opportunities, and share information with others. Some examples of such societies are listed here:

- Discipline-specific
  - American Society of Civil Engineers (ASCE): www.asce.org
  - American Society of Mechanical Engineers (ASME): www.asme.org
  - Institute for Electrical and Electronics Engineers (IEEE): www.ieee.org
  - Society of Automotive Engineers (SAE): www.sae.org
  - American Institute of Chemical Engineers (AIChE): www.aiche.org
  - Society of Manufacturing Engineers (SME): www.sme.org
  - American Academy of Environmental Engineers and Scientists (AAEES): www.aaees.org
  - American Society of Agricultural and Biological Engineers (ASABE): www.asabe.org

- General interest
  - Society of Women Engineers (SWE): www.swe.org
  - Society of Hispanic Professional Engineers (SHPE): www.shpe.org
  - National Society of Black Engineers (NSBE): www.nsbe.org
  - National Society of Professional Engineers (NSPE): www.nspe.org

Engineers will be called on to deal with the challenges that modern society faces. In particular, engineers will be called on to work on sustainable energy initiatives, environmental issues, improving infrastructure, and innovating in the face of global competition. Engineers and engineering technologists are at the forefront of improving the quality of life for all people by using their skills and creativity. You will be called on to use finite natural resources in the most economical way in order to keep humankind moving forward. Are you ready to be part of the solution?

Preparation to Start Your Engineering Studies

To be part of the solution to the needs and aspirations of humankind, engineering students will need to successfully complete their general education, math, science, and engineering coursework. For many students, it is a huge leap from high school to being a student at the university level. Other students will be coming to a university from another career, a junior college, or from the military. All students can benefit from a few tips that the authors will share in this section.

Establish a routine and stick to it. Make sure to attend all classes and labs in a well-established schedule. It may serve you well to allocate set times for reading, doing homework, and writing lab reports. Faculty will tell you that students do better if they make sure to come to class and lab ready to take notes and to participate. Some classes and labs will require that materials be reviewed before attendance. If so, make sure to do that. In addition to class and lab attendance, a student should plan on about two hours per week per credit hour of class or lab to read required material and prepare assignments. For example, if the student takes Chemistry, plan on about eight hours per week to complete homework and lab assignments. There is a reason that 12 semester credit hours (SCH) is a "full load." Consider a student taking 15 SCH. That's 15 hours in class and 30 hours outside class, doing the work necessary to meet the course objectives. That is 45 total hours spent during the week to complete assignments and to study.

Ask questions. Many times, faculty and instructors will ask for questions and there will be silence in a lab or class. Students are often shy about asking questions because they may feel embarrassed, but it is better to ask than to struggle without knowing the answer. Chances are that if you have a question, other students have the same or similar questions. If you are afraid to ask in front of the class, then visit with the professor or instructor before or after the class or lab or during office hours.

Go to tutoring and form study groups. Does your university have tutoring for math, chemistry, physics, English, and more? Your tuition and fees pay for tutoring, so use it. Do a quick online search to find the hours of operation and location of the tutoring services that you need, or ask your faculty or instructor. Also, get to know your fellow students and form study groups. Many students remark that they learn class and lab content better if they practice at tutoring and/or in their study groups.

Do your homework and labs. These assignments are designed as practice and to ensure that you know the material before you take midterm and final exams. Your faculty and instructors will tell you that there is a strong correlation between completed assignments and test scores. Those that do not complete assignments in a timely manner usually do not successfully complete their coursework.

Study for exams. As just mentioned, there is a strong correlation between completion of assignments and course performance. Make sure to review the assignments before going to an exam. Also, outlining the concepts to be covered on the exam can help a student focus on what he or she must study. Your assignments are typically a small percentage of the problems available in your textbook. You might consider working problems beyond those required as practice for your exams.

Make sure to rest and get involved in fun activities. You will need some balance in your school and personal life. Many students get involved in engineering technical societies, other campus groups, intramural sports, exercise, campus ministries, and more. You will come back refreshed for studies. Proper rest, nutrition, and attending to personal business outside of study time are all recommended so that you can devote the necessary time and attention to your coursework. Also know that there are a wide variety of student services available to help you attend to body, mind, and spirit as needed.

Summary

Engineering involves the use of mathematics and science to produce useful goods and services that are required in modern society. Engineers are professionals, similar to doctors and lawyers, due to the fact that these careers require specialized education and are subject to education and licensing requirements.

There are many engineering disciplines to choose from, such as mechanical, civil, electrical, computer science (within engineering), chemical, biomedical, industrial, aerospace, environmental, nuclear, and other areas of engineering specialization.

Theory mastered in mathematics and science enables engineers to design and create new engineered systems. Engineering technologists work with engineers by applying engineering principles to improve and implement engineered systems. Engineering technology is a degree program (bachelor's and master's level) at some universities and an option within industrial engineering in others.

Engineering is a profession that is regulated through its educational and licensing process. Engineers study mathematics, science, university core requirements, engineering core subjects, and engineering discipline-specific coursework. Engineers must continually update their education after graduation due to technological advances.

Engineers can join engineering technical organizations in their respective disciplines and/or those with a general interest. These organizations help to share specifications and technical information, as well as serve networking and other career-related goals.

Engineering students need to prepare for study at the university level. Keeping a routine, asking questions, going to tutoring and study groups, completing homework and lab assignments, studying for exams, and attending to personal needs are all important in ensuring academic success. When in doubt, seek out help and wise counsel, and you will thrive.

Assignment

The first question is to be typewritten using a memorandum format, as shown on the next page. The remaining questions are to be typewritten using a word processor. This is an individual assignment for each student to complete.

1. Using the memo format provided at the end of this chapter, write a one-page summary memo. You may use the built-in memo templates in Microsoft Word or create your own, patterned after the provided memo format. Each student will address the following:

   - Include one paragraph to briefly describe yourself, such as your hometown, the school subjects you like, your interest in engineering, your hobbies, your current (or recent) job, and so forth.
   - Include another paragraph to describe what you hope to get out of this class and your future goals, such as what you may do when you graduate with your degree.

2. Complete the following statement, "I want to become an engineer to design ______," by writing a few well-constructed sentences. Think about what you would like to do as an engineer and what motivates you.

3. Identify the name of a prominent engineer (past or present) and briefly describe in a few well-constructed sentences what he or she did or does as part of his or her job.

4. Our modern society is in a constant state of change in regard to technological and social issues. Briefly describe with a few well-constructed sentences one of the challenges that society faces and how engineers might be able to assist with that challenge.

5. Identify five character traits of yourself that you feel will be helpful in an engineering career. You may use a bulleted list with a short description OR write full sentences describing how each trait may be useful in your future career.

6. To better understand your own learning style, take the Myers-Briggs personality test (the free test is located online at https://www.16personalities.com), which states that there are 16 basic personality types. Descriptions for the personality types can readily be found on this website, as well as on Wikipedia and other online resources. All you need to report is your four-letter type and briefly mention whether you agree or disagree with the result of the test. Note: It is possible to be "halfway between" types; this is fine, just report the type(s) and whether one or more of those types is indicative of your personality as you perceive it.

7. Go to your university's website and find the curriculum guide under your catalog year for your chosen or preferred engineering or engineering technology degree. Prepare a schedule of which courses you plan to take during each semester until you graduate. Consider potential mitigating factors such as work, family, and extracurricular activities in your plan.

Memorandum

TO:	Dr. Just N. Case, PE
FROM:	Ima Friend *IF,* Russ T. Steele *RTS*, and Jane A. Good-Student *JAGS* **[Note that *initials are <u>handwritten in ink</u>* to signify acceptance of content. Do *not* type in the initials.]**
CC:	Dr. Bee A. Genius **[Note: Omit this block if not used, including removal of the blank line.]**
DATE:	Mon 26 Aug 2019
SUBJECT:	ENGR 1301 Class Discussion of Sample Memorandum

The class discussed the essential parts of a memorandum. In particular, the memorandum consists of a heading with *from, to, cc* (optional courtesy copy for other recipients), *subject,* and *date* fields. The body is a concise summary of the major discussion or findings, such as the results of a laboratory test or business that was conducted at a meeting. The *attachments* section is optional and contains any relevant supporting documentation, such as a graph, a photograph or diagram, a table, or anything else that is helpful in determining the reasoning behind the discussion or findings presented in the memorandum.

The memorandum should be a complete, yet brief, communication. While adhering to this format, the memo should introduce the topic or question and the purpose for it. If appropriate, a technical approach should be clearly described. Finally, the question should be answered. For communication of a technical nature, this would also include specific findings and results. A few key points about memoranda are that the length should be no more than two pages and that major word processing software programs, such as Microsoft Word, have templates that can be used to make professional-looking memoranda. The author(s) should always review the content of the memorandum for its factual content and use spelling- and grammar-checking to ensure that the content can be clearly understood. The author(s) should write from an objective standpoint (do not use first-person pronouns such as *I, we,* and *us* unless there is a specific reason to do so) and avoid slang terms (ex: *a lot, gonna, sort of,* etc.) and contractions. Finally, the author(s) of the memorandum should initial in ink by their name on the memorandum to signify acceptance of the final content, since a memorandum can be typed by another person, such as a secretary or a project group member.

Attachments: **[Call attention to attached documents if included; otherwise, omit this block.]**

- Graph
- Diagram
- Spreadsheet Calculations
- Raw Data (etc., as appropriate)

Engineering Measurements and Significant Figures

Objectives

- Determine the number of *significant figures* in a measurement.
- Perform numerical calculations with measured quantities using the correct number of significant digits.

Engineering Measurements

Engineers utilize numbers for most analysis and design functions. They must measure physical quantities and express them in numerical form. They also need to be confident that the numbers they are using are reasonable, since they will make decisions based on these values and the resulting calculations. If a number is only known to 3 digits of accuracy, it should not be reported to 8 digits just because 8 digits appear on the calculator. Readers of your work expect that the number reported reflects how accurately that value is known. All measurements have a number and unit attached. The unit describes the physical quantity in question.

Exact Numbers

- Integers: There are 72 eggs in six one-dozen cartons of eggs.
- By definition: 1 inch = 25.4 mm (exact).

Approximate Numbers

- Measurements: It is 150 miles from campus to my parents' house.
- Irrational numbers: $\pi = 3.141592$

Any physical measurement that is not a countable number is *approximate*. Engineers must ask two questions in connection with approximate numbers: 1) what error is present, and 2) how can the quality of a measurement be assessed?

EXAMPLE #1

A steel bar that is exactly 2.6524 inches long is measured with a ruler marked in 0.1-inch increments. What is the appropriate length to report of the bar as measured by the ruler?

Generally, we can estimate a measurement with this ruler to the nearest 0.05 in. So if we estimate the length as 2.65 inches, we could note this as 2.65 ± 0.05 inches. The third digit is questionable.

Engineers need to be careful in presenting numerical data and results, regardless of language and cultural barriers. Be aware that many countries use a comma where it is common US practice to use a period to denote a decimal. To avoid confusion, it is better to use scientific notation or prefixes, especially when your audience includes people from other countries.

Zeroes, Decimals, and Commas

a. **ALWAYS place a ZERO in front of a number less than one**. Examples: 0.345 and 0.00125 (*NOT* .345 and .00125, since it is very easy to overlook the decimal point).
b. A space may be used instead of a comma to denote orders of three: Examples: 4 567.8 (versus 4567.8 or 4,567.8) and 0.678 91 (versus 0.67891)

Scientific Notation and Prefixes

a. Examples: 4.5678×10^3 and 6.7891×10^{-1}
b. Examples with Prefixes (with list of prefixes shown below):
   1. Length: 25 760 m = 2.576×10^4 m = 25.76 km
   2. Current: 0.00015 A = 1.5×10^{-4} A = 0.15 mA or 150 mA
   3. Mass: 1 000 000 g = 1×10^6 g = 1 Mg
   4. Time: 0.00001 s = 1×10^{-5} s = 10 ms
   5. Data Storage: 1,000,000,000 bytes = 1×10^9 bytes = 1 Gigabyte (GB)

Significant Figures

Significant figures are any digits aside from a single zero before a decimal point or those zeroes without a nonzero to their left. They are also known to students as *significant digits, sig. figs., sig. digs.,* and *s.f.*

Examples (Significant Figures in Parentheses)

a. 34 (2)
b. 0.180 (3)

   c. 1.24×10^6 (3)
   d. 0.014 (2)
   e. 160 (2)
   f. 160. (3)

There are additional guidelines for the use of significant figures. Sometimes you will not be sure how many significant figures are present, thus necessitating engineering judgment. Generally, the last digit of a measurement is an estimate and is assumed to be significant for calculation purposes. Computers and calculators retain all digits. Make sure to report the final answer in terms of the maximum number of significant figures based on mathematical rules described as follows.

Rounding

Rounding requires an increase in the last kept digit by 1 if the last figure dropped is 5 or greater.

   a. 827.48 becomes 827.5 if using 4 significant figures.
   b. 827.48 becomes 827 if using 3 significant figures.

Multiplication and Division

Multiplication and division requires that the number of significant figures in the answer equal the least number of significant figures in any number used.

   a. (2.43)(17.675) = 42.95025 if the two numbers multiplied are exact; otherwise, round the result to 43.0 based on 3 significant figures in 2.43.
   b. (1.2 ft)(4.67 ft)(2.341 ft) = 13.119 ft^3; use 13 ft^3 based on 2 significant figures.
   c. (2.479 h)(60 min/h) = 148.74 min; use 148.7 min, since the 60 of the conversion factor is **exact** and is considered to have infinite significant figures.
   d. $(4.00 \times 10^2$ kg)(2.2046 lb$_m$/kg) = 881.84 lb$_m$; use 882 lb$_m$, since the conversion factor is inexact but has more significant figures than the number being converted, which will govern.

Addition and Subtraction

Addition and subtraction can be done in columns or a horizontal line according to the following rules:

- If arranging in columns
  - Identify the column with the rightmost sig. fig. in each number.
  - Perform the addition and/or subtraction.
  - Round answer so that the rightmost sig. fig. in answer occurs in the leftmost column.

- If arranging in a horizontal line, round the final answer according to the least precise number in the calculation.

$$\begin{array}{r} 14.12 \\ +\,2117.9 \\ \hline 2132.02 \end{array} \rightarrow 2132.0 \qquad \begin{array}{r} 91.71 \\ -\,0.928 \\ \hline 90.782 \end{array} \rightarrow 90.78$$

Combined Operations

When combining numerical operations, attend to the parentheses first, followed by multiplication/division and then addition/subtraction. When carrying the calculations internally in a calculator or computer, the entire calculation is performed and then reported to a reasonable number of sig. figs. (usually 3 to 4).

$$(91.71 - 0.928)(2.24) = 203.35168 \rightarrow 203 \text{ (based on the final multiplication)}$$
$$(4.21 \times 97.8) + 5.5 = 417.238 \rightarrow 417.2 \text{ (based on the final addition)}$$

Summary

Engineers utilize numbers for most analysis and design functions. They must measure physical quantities and express them in numerical form. These numbers have a specific number of significant figures in accordance with the resolution of the measuring device used to determine them. Scientific notation and the use of prefixes help to remove some of the ambiguity associated with numerical values.

There are specific rules for multiplication/division, addition/subtraction, and mixed operations in regard to significant figures.

Assignment

For the following questions, please show your calculations on engineering paper. This is an individual assignment for each student to complete.

1. Report the correct number of significant figures for the following. NOTE: Commas OR spaces maybe be indicated below for powers of three and have no effect on the determination of number of significant figures.

   a. 82,580
   b. 8,258.0
   c. 0.0242
   d. 180 000 000
   e. 1.2×10^4
   f. 0.60801
   g. 3450.10
   h. 2.224×10^{-3}
   i. 90,000,000.0
   j. 4.5678×10^4

2. Report the correct number of significant figures for the following numerical quantities OR report whether they are *exact* quantities for conversions.

   a. 3.7890 ft
   b. 0.100 in

    c. 9.91 lb_m

    d. 25.4 mm/in

    e. 2.2046 $\text{kg}/\text{lb}_\text{m}$

    f. 3600 s/h (seconds per hour)

    g. 60 s/min (seconds per minute)

    h. 15 ft^3

    i. 121.4 km/h

    j. 9.81 m/s^2

3. Perform the following computations, with the correct number of significant figures reported in the FINAL answer.

    a. 56.2×472.08

    b. $3.624 - 2.4200$

    c. $(6.2318)(2.356 \times 10^6)$

    d. $0.32/0.64$

    e. $22.87/4.02$

    f. $(2.24 \times 10^{-3})(81.2 \times 10^6)$

    g. $1824.5/0.180001$

    h. $800.01 \times (4.4 \times 10^2)$

    i. $(20.1 + 14.47 - 0.081)(14.2)$

    j. $(12.77)(3.12)(0.8771)$

4. Perform the following conversion calculations, with the correct number of significant figures reported in the FINAL answer.

    a. 668 ft to miles if 1 mi = 5280 ft (exact)

    b. 7×10^{40} atoms to moles if 1 mol = 6.022×10^{23} atoms (inexact)

    c. 235.3 kg to lb_m if 1 kg = 2.205 lb_m (inexact)

    d. 10.5 weeks to minutes (think about the conversions and whether they are exact)

    e. 12.25 cubic meters to cubic feet if 1 m = 3.2808 ft (inexact)

Engineering Units, Conversions, and Dimensional Analysis

Objectives

- Identify physical quantities in terms of dimensions and units.
- Differentiate between fundamental and derived dimensions.
- Understand the use of non-SI dimensional systems.
- Recognize base, supplementary, and derived SI units.
- Apply the appropriate SI symbols and prefixes.
- Describe the relationship between *USCS* and *SI* systems.
- Convert units from one system to another.
- Use knowledge of dimensions and units to solve engineering problems.

Units

Before world trade was common, countries had their own systems of measurement. As commerce and scientific developments occurred, it became clear that having a common measurement system would allow people to share and benefit in regard to goods and knowledge production. First came procedures for unit systems conversion, and then, later, the development of a truly global standard, the SI standard.

The United States is in the process of metrication, that is, conversion to the global *SI* (International System, a.k.a. metric) standard. Congress considered adoption of the metric system in 1850 and legalized its use in 1866. By the Treaty of the Meter (1875), the United States and 16 other countries established an international agency to oversee a metric standard (base units: meter, kilogram, second). In 1960, the metric system was updated and replaced by the SI standard. The

United States is about halfway through the SI adoption process in an inconsistent manner, as it is not mandatory. Other English-speaking countries, such as the United Kingdom, Canada, Australia, and Ireland, have largely gone through the metrication process, with some customary units still in use.

At the legislative level, the United States has enacted various laws and policies in regard to the metric system:

- Metric Conversion Act (1975): declared SI "the preferred system of weights and measures for United States trade and commerce," but permitted the use of *US Customary Units (USCS)* in nonbusiness activities. Note: USCS measurements are legally defined in terms of SI units. For example, 1 inch is legally defined as 25.4 mm.
- Metric Education Act (1978): provision to educate school aged children in usage of SI units.
- The Omnibus Trade and Competitiveness Act (1988, expired): amended the Metric Conversion Act to require each federal agency to use SI by fiscal year 1992.
- The Department of Transportation planned to require metric units by 2000, but this plan was canceled by the 1998 highway spending bill. Many states had built and continue to build transportation facilities in metric units as a result. Some states had even replaced mileposts with kilometer posts on highways but then changed them back!

NASA learned a big lesson in unit conversions with the case of the Mars Climate Orbiter. Its use of two different systems was the root cause of the loss of the Mars Climate Orbiter in 1998. NASA specified metric units in the contract, and NASA and other organizations worked in metric units, but one subcontractor, Lockheed Martin, provided thruster performance data to the team in lb_f*s instead of N*s. The spacecraft was intended to orbit Mars at about 150 kilometers (93 mi) altitude, but the incorrect data meant that it probably descended instead to about 57 kilometers (35 mi), burning up in the Martian atmosphere. A $328 million satellite was lost due to a simple mistake!

There are two types of dimensional systems. Both use the same type of length and time units. The difference is the use of mass versus force, where force is defined as mass times acceleration (usually due to the effect of gravity on earth).

- Absolute system:
    - Independent of gravity
    - Base quantities: length, time, mass

- Gravitational system:
    - Widely used by many engineering branches (esp. civil, mechanical, electrical, environmental, etc.)
    - Base quantities: length, time, force

TABLE 3.1 **The Seven SI Base Units**

Quantity	Name	Symbol
Length	meter	m
Time	second	s
Mass	kilogram	kg
Temperature	kelvin	K
Electric current	ampere	A
Luminosity	candela	cd
Amount of chemical substance	mole	mol
Plane angle	radian	rad

TABLE 3.2 **Select Derived SI Units**

Quantity	Name	Symbol	Base Units
Frequency	hertz	Hz	s^{-1}
Force	newton	N	$kg \cdot m \cdot s^{-2}$
Pressure	pascal	Pa	$N \cdot m^{-2} = kg \cdot m^{-1} \cdot s^{-2}$
Work, Energy, and Heat	joule	J	$kg \cdot m^2 \cdot s^{-2} = N \cdot m$
Electric Charge	coulomb	C	$s \cdot A$
Electric Potential	volt	V	$kg \cdot m^2 \cdot s^{-3} \cdot A^{-1}$
Electric Resistance	ohm	Ω	$kg \cdot m^2 \cdot s^{-3} \cdot A^{-2}$
Power	watt	W	$kg \cdot m^2 \cdot s^{-3}$

TABLE 3.3 **SI (Metric) Prefixes**

Multiplier	Prefix Name	Symbol
10^{18}	exa	E
10^{15}	peta	P
10^{12}	tera	T
10^{9}	giga	G
10^{6}	mega	M
10^{3}	kilo	k
10^{2}	hecto	h
10^{1}	deka	da
10^{-1}	deci	d
10^{-2}	centi	c
10^{-3}	milli	m
10^{-6}	micro	μ
10^{-9}	nano	n
10^{-12}	pico	p
10^{-15}	femto	f
10^{-18}	atto	a

SI Unit Usage Rules

- SI symbol and naming rules:
  1. Do not place a period after the symbol.
  2. Symbols are lowercase unless derived from someone's name. Ex: m, kg, s, N, A, Hz, Pa, W, and J.
  3. Always use the symbol, not an abbreviation. Ex: s (not sec).
  4. Never add *s* to the symbol for a plural. Ex: This box has a mass of 60 kg (not kgs).
  5. Add space between symbol and number: Ex: 5.5 m (not 5.5 m).
  6. No space between prefix and symbol. Ex: MPa, kg, mm.
  7. Unit names are written out in lowercase, except at the beginning of a sentence. Ex: pascal, newton, hertz, watt, and joule.
  8. Plurals are used when writing out unit names. Ex: fifteen newtons, five grams. Exception: twenty hertz, four lux.
  9. Do not place a hyphen between the prefix and symbol.
  10. Do not mix numerals and written names, or vice versa. Ex: Nine meters or 9 m but not 9 meters or nine m

- SI derived unit examples:
  1. newton meter or newton-meter
  2. meter per second (not meter/second)
  3. millimeter squared or square millimeter or mm^2
  4. $N \cdot m$ or N^*m
  5. m/s or $m \cdot s^1$

There are two variants of older measurement systems commonly used in the United States. The first is the US Customary system (USCS), which has the fundamental units of lb (force), ft (length), s (time), and slug (mass). The second, the more dominant *Engineering System*, has the fundamental units of lb_f (force), ft (length), s (time), and lb_m (mass). This is based on $g_c = 32.1740$ $lb_m \cdot ft/lb_f \cdot s^2$ (earth's gravity at sea level) as a proportionality constant.

There are two types of temperature scales for USCS and SI systems, one based on freezing/boiling of water and one based on absolute zero. For USCS and SI, these are Fahrenheit (°F) and Rankine (°R), and Celsius (°C) and Kelvin (K), respectively. The conversions for each scale is given below, where the letter T refers to temperature.

- $T(°C) = T(K) - 273.15$
- $T(°F) = T(°R) - 459.67$
- $T(°R) = 1.8T(K)$
- $T(°F) = 1.8T(°C) + 32°F$

TABLE 3.4 **A Comparison of the Four Temperature Scales for Water at Sea Level**

Scale	Absolute Zero	Freezing	Boiling
Kelvin	0 K	273.15 K	373.15 K
Celsius	–273.15 °C	0 °C	100 °C
Rankine	0 °R	491.67 °R	671.67 °R
Fahrenheit	–459.67 °F	32 °F	212 °F

Conversions and Dimensional Analysis

Dimensional analysis is one of the most useful problem-solving skills that you will learn in this class. It is all about conversion—converting one thing to another. This is a tool that you will use in every engineering class that you have from this point forward and, more importantly, in everyday life on and off the job. Dimensional analysis is about applied math, not about numbers in the abstract. We're talking about measurable quantities that you can count. Anything you measure will have a number with some sort of "unit of measure" (the dimension) attached. A unit could be miles, gallons, miles per second, peas per pod, or pizza slices per person.

EXAMPLE #1

How many seconds are in a day?

First, do not panic. If you have no idea what the answer is or how to come up with an answer, that is fine; you are not supposed to know. You are not going to solve the entire problem at once. What you are going to do is break the problem down into several small problems that you can solve. (As problems become more complicated in your curriculum, you will find this a very helpful strategy.)

Here is your first problem:

1. Ask yourself, "What units of measure do I want to know or have in the answer?" In this problem, you want to know "seconds in a day." After you figure out what units you want to know, translate the words into math. Math is a sort of shorthand language for writing about numbers of things. If you can rephrase what you want to know using the word *per*, which means "divided by," then that is a step in the right direction, so rephrase "seconds in a day" to "seconds per day." In math terms, what you want to know is:

$$\frac{\text{seconds}}{\text{day}} \text{ or } \frac{\text{s}}{\text{day}}$$

2. Ask, "What do I know?" What do you know about how "seconds" or "days" relate to other units of time measure? You know that there are 60 seconds in a minute. You also know that in 1 minute there are 60 seconds. These are two ways of saying the same thing. You know

that there are 24 hours in a day (and in one day there are 24 hours). If you could now connect "hours" and "minutes" together you would have a sort of bridge that would connect "seconds" to "days" (seconds to minutes to hours to days). The connection you need, of course, is that there are 60 minutes in an hour (and in one hour there are 60 minutes). When you have this kind of connection between units, then you know enough to solve the problem, but first translate what you know into math terms that you can use when solving the problem. If in doubt, write it out:

$$\frac{60\text{ s}}{1\text{ min}} \quad \frac{60\text{ min}}{1\text{ hr}} \quad \frac{1\text{ hr}}{60\text{ min}} \quad \frac{1\text{ day}}{24\text{ hr}} \quad \frac{24\text{ hr}}{1\text{ day}}$$

All of these statements, or conversion factors, are true or equivalent (60 seconds = 1 minute). All you need to do now is pick from these statements the ones that you actually need for this problem, so… .

3. Ask, "From all the factors I know, what do I need to know?"

Remember that you want to know:

$$\frac{\text{seconds}}{\text{day}} \quad \text{or} \quad \frac{\text{s}}{\text{day}}$$

So pick from the things you know a factor that has seconds on top or day(s) on the bottom. You could pick either of the following two factors as your "starting factor:"

$$\frac{60\text{ s}}{1\text{ min}} \quad \text{or} \quad \frac{24\text{ hr}}{1\text{ day}}$$

Write down your starting factor (say you pick 60 seconds per 1 minute):

$$\frac{60\text{ s}}{1\text{ min}}$$

Now the trick is to pick from the other things you know another factor that will cancel out the unit you do not want. You start with "seconds" on top. You want "seconds" on top in your answer, so forget about the seconds—they're okay. The problem is that you have "minutes" on the bottom, but you want "days." You need to get rid of the minutes. You cancel minutes out by picking a factor that has minutes on top. With minutes on top and bottom, the minutes will cancel out. So you need to pick 60 minutes per 1 hour as the next factor because it has minutes on top:

$$\frac{60\text{ s}}{1\ \cancel{\text{min}}} \ \bigg| \ \frac{60\ \cancel{\text{min}}}{1\text{ hr}}$$

You now have seconds per hour, since the minutes have cancelled out, but you want seconds per day, so you need to pick a factor that cancels out hours:

$$\frac{60\,\text{s}}{1\,\text{min}} \left| \frac{60\,\text{min}}{1\,\text{hr}} \right| \frac{24\,\text{hr}}{1\,\text{day}}$$

4. Solve it. When you have cancelled out the units you don't want and are left only with the units you do want, then you know it is time to multiply all the top numbers together and divide by all the bottom numbers.

$$\frac{60\,\text{s}}{1\,\text{min}} \left| \frac{60\,\text{min}}{1\,\text{hr}} \right| \frac{24\,\text{hr}}{1\,\text{day}} = \frac{86400\,\text{s}}{1\,\text{day}}$$

In this case, you just need to multiply 60 × 60 × 24 to get the answer; there are 86,400 seconds in a day.

Here is how this problem might look if it were written on the board:

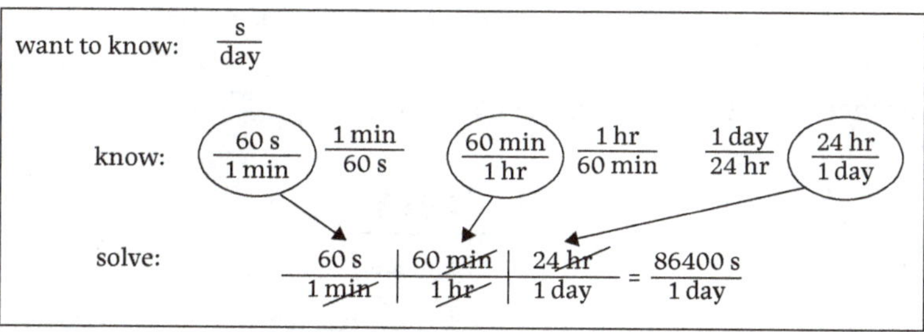

Remember that you do not need to worry about the actual numbers until the very end. Just focus on the units. Plug in conversion factors that cancel out the units you do not want until you end up with the units you do want. Only then do you need to worry about doing the arithmetic. If you set up the bridge so that the units work out, then, unless you push the wrong button on your calculator, you WILL get the right answer every time.

EXAMPLE #2

How many hours are in a year? Go through the steps:

What do you want to know? This one is easy: hours in a year or, better, hours per year

Translate into math terms:

$$\frac{\text{hours}}{\text{year}} \quad \text{or} \quad \frac{\text{hr}}{\text{year}}$$

What do you know that you might need to use? You know that there are 24 hours in a day, 7 days in a week, 4 weeks in a month, 12 months in a year, and about 365 days in a year. The converse, of course, is also true: In one day there are 24 hours, in one week there are 7 days, in one month there are 4 weeks, in one year there are 12 months, and in one year there is about 365 days (actually closer to 365.25 days).

Translate into math terms—writing them all down can't hurt:

$$\frac{24\ \text{hr}}{1\ \text{day}} \quad \frac{7\ \text{days}}{1\ \text{week}} \quad \frac{4\ \text{weeks}}{1\ \text{month}} \quad \frac{12\ \text{months}}{1\ \text{year}} \quad \frac{365\ \text{days}}{1\ \text{year}}$$

$$\frac{1\ \text{day}}{24\ \text{hr}} \quad \frac{1\ \text{week}}{7\ \text{days}} \quad \frac{1\ \text{month}}{4\ \text{weeks}} \quad \frac{1\ \text{year}}{12\ \text{months}} \quad \frac{1\ \text{year}}{365\ \text{days}}$$

Pick a starting factor. Since you want hours on top (or years on the bottom), you could start with 24 hours per day:

$$\frac{24\ \text{hr}}{1\ \text{day}}\bigg|$$

Now pick whatever factors you need to cancel out day(s):

$$\frac{24\ \text{hr}}{1\ \cancel{\text{day}}}\bigg|\frac{7\ \cancel{\text{days}}}{1\ \text{week}}$$

Keep picking factors that cancel out what you don't want until you end up with the units you do want:

$$\frac{24\ \text{hr}}{1\ \cancel{\text{day}}}\bigg|\frac{7\ \cancel{\text{days}}}{1\ \cancel{\text{week}}}\bigg|\frac{4\ \cancel{\text{weeks}}}{1\ \cancel{\text{month}}}\bigg|\frac{12\ \cancel{\text{months}}}{1\ \text{year}}$$

Do the math. You have all the units cancelled out except hours per year, which is what you want, so when you do the math, you know you'll get the right answer:

$$\frac{24\ \text{hr}}{1\ \cancel{\text{day}}}\bigg|\frac{7\ \cancel{\text{days}}}{1\ \cancel{\text{week}}}\bigg|\frac{4\ \cancel{\text{weeks}}}{1\ \cancel{\text{month}}}\bigg|\frac{12\ \cancel{\text{months}}}{1\ \text{year}} = \frac{8064\ \text{hr}}{\text{year}}$$

But wait, why not go from "hours" to "year" using the fact that there are 365 days per year?

$$\frac{24\ \text{hr}}{1\ \cancel{\text{day}}}\bigg|\frac{365\ \cancel{\text{days}}}{1\ \text{year}} = \frac{8760\ \text{hr}}{\text{year}}$$

There is a 696-hour difference between the two answers! How can this be? Exact answers can only be obtained if you use conversion factors that are exactly correct. In this case, there are actually more than 4 weeks in a month (about 4.35 weeks per month). Since there is a bit more than 365 days in a year (365.25 days), a more accurate answer still, to the nearest hour, would be that there are 8,766 hours in a year. The point of this example is that an answer can only be as accurate as the conversion factors used.

EXAMPLE #3

Sometimes both the top and bottom units need to be converted:

If you are going 50 miles per hour, how many feet per second are you traveling?

If you were to do this one on the blackboard, it might look something like this:

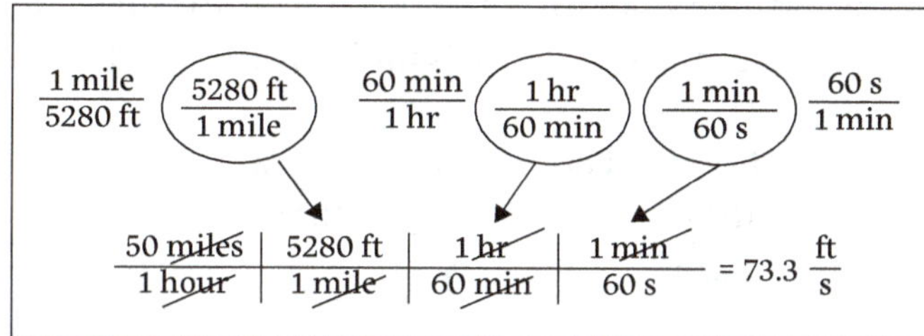

You want your answer to be in feet per second. You are given 50 miles per hour. Normally you can use any value given by the problem as your starting factor. One thing you know, then, is given. The other things you just know or have to look up in a conversion table. Although every conversion factor can be written two ways, you really only need to write each one way. That's because you know you can always just flip it over and then use it. If you have written 60 min/1 hr, then to solve this problem, you would just flip the 60 min/hour factor over. With practice, you won't even need to write down what you know, you'll just write down the last part, do the math, and get the right answer.

EXAMPLE #4

You have come down with a bad case of the geebies, but fortunately, your grandmother knows how to cure the geebies. She sends you an eyedropper bottle labeled:

Take 1 drop per 10 lb of body weight per day divided into 4 doses until the geebies are gone.

This problem is a bit more challenging, but <u>do not panic</u>. Break the problem down into a bunch of small problems, and tackle each one by one.

What do you want to know? In order to take one dose 4 times a day, you need to know how many drops to take per dose.

Translated into math terms, you want the answer to be in:

$$\frac{\text{drops}}{\text{dose}}$$

What do you know? Well, you know that you weigh 160 pounds. You know that you need to take 4 doses per day (implied). You know that you need to take 1 drop per 10-lb body weight per day.

Translate this into math terms:

$$\frac{160\,\text{lb}}{}\quad \frac{4\,\text{doses}}{1\,\text{day}}\quad \frac{1\,\text{drop}}{\dfrac{10\,\text{lb}}{1\,\text{day}}}$$

The problem now is the last factor. What can you do with such an odd factor? You rewrite it so it is in a different form that you can use. What you do is multiply the middle term by the bottom term:

$$\frac{\dfrac{1\,\text{drop}}{10\,\text{lbs}}}{1\,\text{day}} = \frac{1\,\text{drop}}{10\,\text{lb}} \times \frac{1}{1\,\text{day}} = \frac{1\,\text{drop}}{10\,\text{lb} \times 1\,\text{day}}$$

So, whenever you have a triple-decker, use this trick to rewrite the factor.

Now you should be able to solve the problem. The one thing you know that is not a conversion factor is that you weigh 160 lb, so use that as your starting factor:

$$\frac{160\,\text{lb}}{}\left|\frac{1\,\text{drop}}{10\,\text{lb}\,|\,1\,\text{day}}\right|\frac{1\,\text{day}}{4\,\text{doses}} = 4\,\frac{\text{drops}}{\text{dose}}$$

If you had wanted to know how many drops per day to take, you would have just left off the last conversion factor, which would give you an answer of 16 drops/day.

EXAMPLE #5

How much bleach would you need to make a quart of 5-percent bleach solution?

In this example, you are not told what answer unit to use, but ounces (oz) would work, since there are 32 oz in a quart. If the only thing you had to measure bleach with was in milliliters, then pick "mL" as your answer unit.

When you are given something like *5 percent* or *5%* by a problem, translate it into math terms you can use. "5%" means 5 per 100 or 5/100, but 5 what? per 100 what? You need to label the numbers appropriately. In this example, you write down "5 oz bleach/100 oz bleach solution" or just "5 oz

B/100 oz bleach solution" as a factor you are given. If you were going for milliliters, then you would use "5 mL B/100 mL bleach solution." The top and bottom units must be the same or equivalent but otherwise can be any units you may need. If you had 5 gal bleach/100 gal bleach solution, then you still have a 5% bleach solution.

Setting up and solving this problem is now easy:

$$\frac{1\ \cancel{qt}\ \cancel{BS}}{} \left| \frac{32\ \cancel{oz}}{1\ \cancel{qt}} \right| \frac{5\ oz\ B}{100\ \cancel{oz}\ \cancel{BS}} = 1.6\ oz\ bleach$$

So, you would add 1.6 oz bleach to a quart measuring cup, and then add water (30.4 oz) to make 32 ounces of 5% bleach solution.

EXAMPLE #6

Your car's gas tank holds 18.6 gallons and is one quarter full. Your car gets 16 miles/gal. You see a sign saying, "Next gas 73 miles." Your often-wrong brother, who is driving, is sure you'll make it without running out of gas. You're not so sure and do some quick figuring:

$$\frac{18.6\ \cancel{gal}}{1\ \cancel{tank}} \left| \frac{1\ \cancel{tank}}{4\ quarter\text{--}tank} \right| \frac{16\ miles}{1\ \cancel{gal}} = \frac{74.4\ miles}{quarter\text{--}tank}$$

"Ah! I knew I was right," declares your brother upon glancing at your calculator. "Your calculations prove it!" Is this a good time to be assertive and demand that he turn around to get gas, or do you conclude that your brother is right for once?

Unless the next stop is Las Vegas and you're feeling really lucky, you should turn around and get some more gas. Your calculator reads "74.4" exactly, but if you believe what it says, then you believe that when the car runs out of gas, it will have gone somewhere between 74.35 miles and 74.45 miles or 74.4 ± 0.05 miles. Obviously the "point four" is meaningless, so you round to 74. Is 74 the right answer? If you think it is, then you think your car will go somewhere between 73.5 miles and 74.5 miles before running out of gas. When you run out of gas, what chance do you think you'll have of having gone 74 ± 0.5 miles? A proverbial "fat chance" would be a good guess.

In addition, when you're given something like a "quarter tank," you should wonder just how accurate such a measurement is. Can you really divide 18.6 by 4 and conclude that there really is 4.65 gallons of gas in the tank? The last time you figured mileage, you came up with 16 miles/gallon, but is the engine still operating as efficiently? Are the tires still properly inflated? Will the next 73 miles be uphill? Do you have a head wind? Surely a calculated estimate of 74 miles is overly precise. A realistic answer, then, might be 74 ± 10 miles. You realize you could run out of gas anywhere between 66 and 84 miles, so you finally and correctly conclude that you have a slightly less than 50/50 chance of running out of gas before reaching the next gas station.

The point of this example is to remind you that dimensional analysis is applied math, not abstract math. The numbers used should describe the real world insofar as possible and indicate no more accuracy than is appropriate. If you overlook this point, you might have a five-mile walk to the next gas station.

EXAMPLE #7

You are throwing a pizza party for 15 people and figure each person might eat 4 slices. How much is the pizza going to cost you? You call up the pizza place and learn that each pizza will cost you $14.78 and will be cut into 12 slices. You tell them you will call back. Do you have enough money? Here is how you figure it out, step by step.

1. Ask yourself, "What do I want to know?" In this case, how much money is the pizza going to cost you, which in math terms is: cost (in dollars) per party, or just $/party. This is your "answer unit." This is what you are looking for.

2. Ask, "What do I know?" Write it all down, everything you know: one pizza will cost you $14.78 (in math terms, 1 pizza/$14.78). You also know that for $14.78 you can buy one pizza ($14.78/1 pizza). It can be important to realize that every conversion factor you know can be written two ways. One of these ways may be needed to solve the problem and the other will not, but in the beginning, you don't know which, so just write them both ways. Continue writing down other things you know. You know, or hope, that only 15 people will be eating pizza (15 persons/1 party), or for this one party, 15 people will come (1 party/15 persons). You also know there will be 12 slices per pizza (12 slices/1 pizza), or that each pizza has 12 slices (1 pizza/12 slices). The last thing you know is that each person gets 4 slices (1 person/4 slices), or that you are buying 4 slices per person (4 slices/person). Math is a language that is much briefer and clearer than English, so writing everything you know in math terms, here is what you might have written down:

$$\frac{1 \text{ pizza}}{\$14.78} \quad \frac{\$14.78}{1 \text{ pizza}} \quad \frac{15 \text{ persons}}{1 \text{ party}} \quad \frac{1 \text{ party}}{15 \text{ persons}}$$

$$\frac{12 \text{ slices}}{1 \text{ pizza}} \quad \frac{1 \text{ pizza}}{12 \text{ slices}} \quad \frac{1 \text{ person}}{4 \text{ slices}} \quad \frac{4 \text{ slices}}{1 \text{ person}}$$

3. Ask, "From all the things above that I know, what do I actually need to know to figure out the problem?"

   Remember that you want to know $/party, so pick one of the things you know that has either dollars on top or party on the bottom. Let's start with $14.78/pizza as the starting factor. Great, you've got dollars on top, but you have pizza on the bottom where you want party.

To get rid of pizza, pick one of the things you know that has pizza on the top. Pizza over pizza cancels out, so you get rid of the pizza.

$$\frac{\$14.78}{1\ \text{pizza}} \left| \frac{1\ \text{pizza}}{12\ \text{slices}} = \frac{\$1.23}{\text{slice}}$$

Okay, you now have dollars per slice, but you want dollars per party, so now what? Easy, just keep picking from the things you know whatever cancels out the units you don't want. The numbers go with the units, but don't worry about numbers, just pay attention to the units. So you pick 4 slices/1 person to get rid of *slices*, then 15 persons/1 party to get rid of *persons*.

$$\frac{\$14.78}{1\ \text{pizza}} \left| \frac{1\ \text{pizza}}{12\ \text{slices}} \right| \frac{4\ \text{slices}}{1\ \text{person}} \left| \frac{15\ \text{persons}}{1\ \text{party}} = \frac{14.78 \times 4 \times 15}{12} = 73.90 = \frac{\$73.90}{\text{party}}$$

Now multiply all the top numbers, and then divide by any bottom numbers to get the right number. Finally, add the units that are left over to the number to get the answer you wanted. Using this method, you can hardly go wrong unless you push the wrong button on your calculator.

By the way, how many pizzas should you order? Figuring this out should be as easy as… .

$$\frac{1\ \text{pizza}}{12\ \text{slices}} \left| \frac{4\ \text{slices}}{1\ \text{person}} \right| \frac{15\ \text{persons}}{1\ \text{party}} = \frac{5\ \text{pizzas}}{\text{party}}$$

EXAMPLE #8

Chemists often use dimensional analysis. Here is a chemistry problem. To solve it, you need to know that, as always, there are 6.02×10^{23} molecules (or atoms) of a substance in 1 mole.

A sample of calcium nitrate, $Ca(NO_3)_2$, with a formula weight of 164 g/mol, has 5.00×10^{25} atoms of oxygen. How many kilograms of $Ca(NO_3)_2$ are present?

Since you want kilograms (kg) in your answer, pick a starting factor with mass (grams, or *g*) on top. Note that in each molecule of calcium nitrate there are two nitrate ions, each having three atoms of oxygen (for a total of 6).

$$\frac{64\ g\ Ca(NO_3)_2}{1\ mol} \left| \frac{1\ mol}{6.02 \times 10^{23}\ \text{molecules}\ Ca(NO_3)_2} \right| \frac{1\ \text{molecule}\ Ca(NO_3)_2}{6\ \text{atoms Oxygen}}$$

$$\frac{5.00 \times 10^{25}\ \text{atoms Oxygen}}{\text{sample}} \left| \frac{1\ kg}{1000\ g} = \frac{2.27\ kg\ Ca(NO_3)_2}{\text{sample}}$$

EXAMPLE #9

Okay, enough easy problems—try something harder. Well, not really harder, just longer. The point of this example is that no matter how ridiculously long your conversion might be, long problems are not really more difficult. If you get the point, then skip this example; otherwise, read on.

At the pizza party, you and two friends decide to go to Mexico City from El Paso, Texas, where you live. You volunteer your car if everyone chips in for gas. Someone asks how much the gas will cost per person on a round trip. Your first step is to call your smarter brother to see if he will figure it out for you. Naturally, he is too busy to bother, but he does tell you that it is 2,015 km to Mexico City, there are 11 cents to the peso, and gas costs 5.8 pesos per liter in Mexico. You know your car gets 21 miles to the gallon, but we still do not have a clue as to how much the trip is going to cost (in dollars) each person in gas ($/person).

1. What do you want to know? $/person round trip—the answer unit(s).

2. What do you know so far? There will be 3 persons going on a round trip (3 persons/1 round trip), or, in the planned round trip, 3 persons will be going (1 round trip/3 persons); it will be a 2,015 km trip one-way (2015 km/one-way trip), or one-way is 2,015 km (one-way trip/2015 km); there are two one-way trips per round trip (2 one-way/round trip); and there are 11 cents per peso (11 cents/1 peso), or one peso is worth 11 cents (1 peso/11 cents). Finally, you know that one liter of gas costs 5.8 pesos (1 liter/ 5.8 pesos), or 5.8 pesos will buy you 1 liter of gas (5.8 pesos/1 liter).

    You know a lot, but still not enough. Knowing the number of miles in a kilometer or liters in a gallon would be nice, but one of your friends recalls that there are 39.37 inches in a meter, and the other is sure that there are 4.9 mL in a teaspoon. This still is not enough. You might need to know that there are 1,000 meters in a kilometer, 1,000 mL in a liter, 100 cents in a dollar, 12 inches in a foot, and 5,280 feet in a mile—but then, you already knew that.
    Almost enough, but how can you get from teaspoons to gallons? Simple, call your mom. She knows that there are 3 teaspoons in a tablespoon, 16 tablespoons in a cup, 2 cups in a pint, 2 pints in a quart, and 4 quarts in a gallon. Wow, that is a lot of things to know, but it should be enough. Write it all down in math terms and see what you have:

$$\frac{1\ \text{round trip}}{3\ \text{persons}} \quad \frac{2015\ \text{km}}{\text{one}-\text{way}} \quad \frac{2\ \text{one}-\text{way}}{1\ \text{round trip}} \quad \frac{11\ \text{cents}}{1\ \text{peso}} \quad \frac{1\ \text{liter}}{5.8\ \text{pesos}} \quad \frac{21\ \text{miles}}{1\ \text{gallon}} \quad \frac{39.37\ \text{in}}{1\ \text{m}} \quad \frac{1000\ \text{m}}{1\ \text{km}}$$

$$\frac{1000\ \text{mL}}{1\ \text{liter}} \quad \frac{12\ \text{in}}{1\ \text{ft}} \quad \frac{5280\ \text{ft}}{1\ \text{mile}} \quad \frac{100\ \text{cents}}{1\ \text{dollar}} \quad \frac{4.9\ \text{mL}}{1\ \text{tsp}} \quad \frac{3\ \text{tsp}}{1\ \text{tbs}} \quad \frac{16\ \text{tbs}}{1\ \text{cup}} \quad \frac{2\ \text{cup}}{1\ \text{pint}} \quad \frac{2\ \text{pints}}{1\ \text{qt}} \quad \frac{4\ \text{qt}}{1\ \text{gal}}$$

3. What do you need to know from the above? If any of the above are upside down from what you end up needing, just turn them over, then use them. This problem looks harder than it is. Since we want to end up with $/person, start with:

$$\frac{1 \text{ round trip}}{3 \text{ persons}} \cdot \frac{2 \text{ one-way}}{1 \text{ round trip}} \cdot \frac{2015 \text{ km}}{\text{one-way}} \cdot \frac{1000 \text{ m}}{1 \text{ km}} \cdot \frac{39.37 \text{ in}}{1 \text{ m}} \cdot \frac{1 \text{ ft}}{12 \text{ in}} \cdot \frac{1 \text{ mile}}{5280 \text{ ft}} \cdot \frac{1 \text{ gal}}{21 \text{ miles}} \cdot \frac{4 \text{ qt}}{1 \text{ gal}}$$

$$\frac{2 \text{ pints}}{1 \text{ qt}} \cdot \frac{2 \text{ cup}}{1 \text{ pint}} \cdot \frac{16 \text{ tbs}}{1 \text{ cup}} \cdot \frac{3 \text{ tsp}}{1 \text{ tbs}} \cdot \frac{4.9 \text{ ml}}{1 \text{ tsp}} \cdot \frac{1 \text{ liter}}{1000 \text{ ml}} \cdot \frac{5.8 \text{ pesos}}{1 \text{ liter}} \cdot \frac{11 \text{ cents}}{1 \text{ peso}} \cdot \frac{1 \text{ dollar}}{100 \text{ cents}} = \frac{\$96}{\text{person}}$$

Even with 18 factors to plug in to get your answer, it is still pretty much a no-brainer, whether you have two or 30 factors. If you know enough conversion factors and set the bridge up correctly to cancel out all unwanted units, then you get the correct answer. You may have to look up a few conversion factors you did not know, but once you do, you are home free. Looking up the number of liters per gallon or miles in a kilometer would have saved quite a few steps, but if you can remember any relationship, you can still figure out your answer. It takes a little longer but really adds nothing to the inherent difficulty.

Summary

There are two primary systems of units currently used in the world: the SI and the USCS. The SI system is a rational system used throughout the world for scientific and business matters with a clearly defined system of base and derived units that utilizes prefixes to denote orders of magnitude. The United States is in the midst of a gradual transition from the older Customary System, which has two primary variants (USCS and the Engineering System). Engineers in the United States need to be proficient in both unit systems, or they may run into serious issues in analysis and design of engineered systems.

When using dimensional analysis to solve an engineering problem, use the following steps:

1. Figure out what answer unit(s) you want to end up with.
2. Write down, in *math terms*, everything you know that relates to the problem. You may need to read the problem several times, rephrasing parts of it, so you can translate everything into math terms. You may need to look up a few conversion factors, but that is simply inconvenient, not difficult.
3. Pick a starting factor. If possible, pick one that already has one of the units you want in the right place. Otherwise, start with something you are given that is not a conversion factor.
4. Plug in conversion factors that allow you to cancel out any units you don't want until you are left with only the units you do want (your answer units).
5. If you cannot solve the problem, pick a different starting factor and start over.
6. Do the math. You may be less apt to make an error if you first multiply all the top numbers, then divide by all the bottom numbers. Then, double-check your calculations. If you make a mistake, it will probably be in hitting the wrong key on the calculator.
7. Ask yourself if the answer seems right or reasonable. If not, recheck everything.

Assignment

1. Using the rules for expressing SI unit *symbols*, correct only those quantities that are incorrectly labeled. If correct, simply indicate a check mark next to the quantity.

   a. 82 amps

   b. 299 degrees Kelvins

   c. 21.1 cm/s*s

   d. 237 pa

   e. 1 MHertz

   f. 44.5 m*m

   g. 6895 n

   h. 2.21 cm's

   i. 14 mA

   j. 27 KW

2. Using the rules for expressing SI unit *symbols*, correct only those quantities that are incorrectly labeled. If correct, simply indicate a check mark next to the quantity.

   a. 25 A's

   b. 71.7 KG

   c. 29 farads

   d. 34 J/s

   e. 7.78 m per sec

   f. 4.95 N

   g. 22 m squared

   h. 34.2 deg C

   i. 9.12 N*m

   j. 55 kilometers

3. 0.56 kg = ? mg

   $$0.56 \text{ kg} \times \underline{\hspace{2cm}} \frac{\text{g}}{\text{kg}} \times \underline{\hspace{2cm}} \frac{\text{mg}}{\text{g}} = \underline{\hspace{2cm}} \text{mg}$$

4. 1.2 ng = ? g

   $$1.2 \text{ ng} \times \underline{\hspace{2cm}} \frac{\text{g}}{\text{ng}} = \underline{\hspace{2cm}} \text{g}$$

5. 2.0 in = ? mm (1 in = 2.54 cm)

   $$2.0 \text{ in} \times \underline{\hspace{2cm}} \frac{\text{cm}}{\text{in}} \times \underline{\hspace{2cm}} \frac{\text{m}}{\text{cm}} \times \underline{\hspace{2cm}} \frac{\text{mm}}{\text{m}} = \underline{\hspace{2cm}} \text{mm}$$

6. 500 ft = ? m

   $$500 \text{ ft} \times \underline{\hspace{2cm}} \frac{\text{in}}{\text{ft}} \times \underline{\hspace{2cm}} \frac{\text{cm}}{\text{in}} \times \underline{\hspace{2cm}} \frac{\text{m}}{\text{cm}} = \underline{\hspace{2cm}} \text{m}$$

7. 10 μL = ? cc (1 mL = 1 cm^3 = 1 cc)

   $$10 \text{ }\mu\text{L} \times \underline{\hspace{2cm}} \frac{\text{L}}{\mu\text{L}} \times \underline{\hspace{2cm}} \frac{\text{mL}}{\text{L}} \times \underline{\hspace{2cm}} \frac{\text{cc}}{\text{mL}} = \underline{\hspace{2cm}} \text{cc}$$

8. 3 wk = ? min

   $$3 \text{ wk} \times \underline{\hspace{2cm}} \times \underline{\hspace{2cm}} \times \underline{\hspace{2cm}} = \underline{\hspace{2cm}} \text{min}$$

9. 50 mL = ? cups (1 L = 4.226 cups)

   $$50 \text{ mL} \times \underline{\hspace{2cm}} \times \underline{\hspace{2cm}} = \underline{\hspace{2cm}} \text{cups}$$

10. 5.33 km = ? dm

    $$5.33 \text{ km} \times \underline{\hspace{2cm}} \times \underline{\hspace{2cm}} = \underline{\hspace{2cm}} \text{dm}$$

11. 123.0 ng = ? Mg

123.0 ng × _________________ × _________________ = _________________ Mg

12. 3 yd = ? in (1 yd = 3 ft)

3 yd × _________________ × _________________ = _________________ in

Estimation of Errors and Approximations

Objectives

- Define *accuracy* and *precision* in measurements.
- Define *systematic* and *random errors* and explain how they occur.
- Solve problems involving approximations in the required data.
- Develop and present calculations using the engineering solution format.

Estimation

Engineers need to be confident that the numbers (estimates) they are using are reasonable. They will make decisions based on these values and the resulting calculations. Thus, two major questions arise: can engineers be 100-percent certain of their numbers, and how do they deal with uncertainty?

The terms *accuracy* and *precision* have very well-defined meanings for mathematicians and engineers alike. *Accuracy* is the measure of nearness of a value to a correct or true value. *Precision* refers to the repeatability of a measurement. These two concepts occur simultaneously in connection with engineering data. Ideally, engineers want to be both accurate and precise with their measurements. The table below illustrates accuracy and precision.

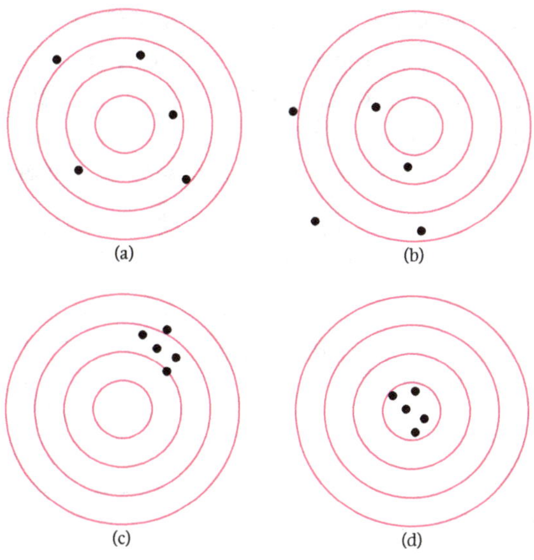

a. accurate, not precise
b. not accurate, not precise
c. not accurate, precise
d. accurate, precise

Errors

Error classification helps the engineer to make wise, informed decisions on the validity of data. Anytime a measurement is taken, the result is compared to a true value, which itself may not be known exactly. Some questions an engineer may ask about the data include the following:

- Did the same person make all of the measurements?
- Was the same measuring device used?
- Was the measuring device correctly graduated?
- Was the measuring device calibrated?
- Did the environment play a role in the measurement?
- Were the proper measurement techniques used?

These questions help in the classification of errors as two major types: *systematic* and *random*. *Systematic errors* are identifiable and correctable, while *random errors* are accidental or not identifiable.

A measuring tape 25.000 ft long is used to measure a distance of 1200 ft. The tape measure has been compared with a standard tape measure to determine its suitability for the task. If the tape is not exactly 25.000 ft, the error will be repeated each of the 48 times it is used to lay out the distance of 1200 ft. This is a systematic error. Other potential systematic errors in the measurement include temperature effects on the measuring tape (which itself is dependent on the error present in a thermometer), the tension applied, and the weight of the tape if suspended in the air.

In the same example of a measuring tape, random error can be introduced if the experiment is not conducted the same every time (differences in reading the distances between marks, not placing the tape in the same starting position each time, differences in sag of the tape, etc.). Random errors can also be introduced when correcting for systematic errors (error reading temperature from a thermometer to account for thermal expansion).

Errors can be minimized. In the case of systematic errors, it is relatively simple to go through and eliminate each correctable error, often by a numerical correction. For random errors, the following suggestions can help: take several readings and average them, improve the quality of the measuring device, and improve the measuring skills of the human operators.

Error propagates in numerical calculations. Two of the most common propagations of error are by summation and by multiplication.

- Error in a sum
  - Let a, b, c, etc., be independent measurements.
  - Then the error in their sum is:

$$E_{sum} = \pm\sqrt{(E_a^2 + E_b^2 + E_c^2 + \ldots)}$$

- Error of a product
  - Let A and B be measurements.
  - Then E_a = error of A and E_b = error of B.
  - Then it can be proved that $E_{product}$ can be found as:

$$E_{product} = \pm\sqrt{(A^2 E_b^2 + B^2 E_a^2)}$$

EXAMPLE #1

a = 100 ft ± 1 ft, b = 130 ft ± 1.5 ft, and c = 200 ft ± 0.7 ft. Find their sum and the error of their sum.

Sum = 100 + 130 + 200 = 330 ft

$E_{sum} = \pm((1)^2 + (1.5)^2 + (0.7)^2)^{0.5} = \pm1.93$ ft

Thus, the length is 330 ft ± 1.93 ft.

EXAMPLE #2

A = 100 ft ± 0.02 ft and B = 200 ft ± 0.03 ft. Find their product and the error of their product.

(A)(B) = (100 ft) (200 ft) = 20,000 ft^2

$E_{product} = ((100 \text{ ft})^2 (0.03 \text{ ft})^2 + (200 \text{ ft})^2 (0.02 \text{ ft})^2)0.5 = 5.00$ ft^2

Thus, the area is 20,000 ft^2 ± 5 ft^2

Approximations

Engineers strive for a high level of precision in their work. Sometimes they need a quick but reasonable estimate using an *approximation*, usually based on past experience with a particular type of problem. Given time and resources, the numbers in the approximation can be refined to get a better answer. A guess is just a guess without reasoning to back it up, so an engineer needs to be reasonably sure about the values used in the approximation. Usually, with practice, your approximation will be within 10–20% of the best analysis that can be done. This is often called a *ball-park approximation* and helps to get you started with a more rigorous calculation process.

Summary

All engineering calculations are estimations and are subject to errors. Errors are either systematic (determinable) or random (indeterminate), but there are methods that can be implemented to minimize the effects of both. The error of a product and of a sum are readily quantifiable, based on the estimated error of each contributing measurement. Engineers can use approximations to get a *ball-park approximation* to begin a more rigorous calculation process; normally an approximation will result in an estimate that is within about 10–20% of the final calculated value.

Assignment

1. Estimate the floor area (ft^2) and room volume (ft^3) of the classroom by

    a. Using the floor/carpet tiles, door frames, and/or ceiling tiles (items that you may be able to estimate the dimensions of).
    b. Using a tape measure.

    Compare the results. How similar are they? (Make sure to sketch the room dimensions for the estimate and for the use of the tape measure.)

2. Using your results (with the tape measure) from Problem #1, estimate the number of the following objects that will fit completely within the room, assuming all furniture and fixtures have been removed. Assume that spherical objects will be in boxes such that they just touch each side of the box they are in. (Partial objects do not count.) Note: 1 in = 25.4 mm, and diameter d = 2r (radius)

    a. A Banker's Box (12 inches wide, 16 inches deep, 10 inches high)
    b. A ping pong ball (40 mm diameter)
    c. A tennis ball (67 mm diameter)
    d. An NBA basketball (29.5 in circumference)

3. Parts b, c, and d in Problem #2 each involve a sphere. If you place a sphere in a cubic box, there will be empty space left over. Determine the percentage of the box taken up by the sphere (it is the same for any sphere).

    a. Write the volume of the box in *general terms* (i.e., symbolically).
    b. Write the volume of the sphere in *general terms* (i.e., symbolically).
    c. Divide the volume of the sphere by that of the box and multiply by 100 to report the percentage.
    d. What does this result mean in regard to shipping spherical objects in cubic boxes? How might you utilize the empty space? (There is no definite answer; just think about what you can do.)

Figure Credits

Fig. 4.1: Source: https://www.open.edu/openlearn/science-maths-technology/mathematics-statistics/using-numbers-and-handling-data/content-section-2.1.

Engineering Problem-Solving

> ## Objectives
> - Describe the steps of the *engineering method*.
> - Use the engineering method as a practical approach to solve engineering problems.

The Engineering Method

Engineers are expected to be able to analyze, understand, and solve a wide variety of problems. Engineering problem-solving combines knowledge, experience, organization, and sometimes a technical artistry. Over time, the field of engineering has developed a reputation for becoming adept at problem-solving in general. This is in part due to the method in which engineers are trained to approach technical problems. This method is a six-step process that is generally considered the *engineering method* to problem-solving.

1. *Recognize and understand the problem.* In the classroom, problems are usually clearly stated, and the student knows what solution to work toward. However, in the real world, defining the problem can be much more abstract. Efforts must be made to completely understand the root of the problem, which factors are significant, and what can be considered to have no influence.
2. *Accumulate data and verify accuracy.* Information pertinent to the problem can be collected through tests and experimentation, or it can come from previous research collected from papers, books, and journals.
3. *Select the appropriate theory or principle.* This can come in the form of simple mathematical equations or vast scientific theories.

4. *Make necessary assumptions.* There are no perfect solutions to real-world problems. In the real world, very little is known with absolute certainty. It is up to the engineer to make assumptions that allow a solution to be obtained without introducing significant error.

5. *Solve the problem.* Combine steps 2, 3, and 4 above to reach a suitable solution.

6. *Verify and check results.* All solutions should be tested using whatever means are available and necessary to verify accuracy. This will sometimes result in an iterative process, wherein the solution is found faulty and the engineer will return to a previous step in the process to refine the method.

Engineering Problem Presentation

The engineering method is the process by which engineers think through problems to achieve a solution. To document this problem-solving process, many engineers have adopted a standard problem presentation format. This format is used worldwide by all types of engineers and will be used for the entirety of your engineering education and career. Information should be included to a level of detail that will allow a knowledgeable engineer to verify or reproduce the calculations and understand their purpose. Note that this format is introduced in your syllabus.

1. *Given:* State clearly but concisely the problem description. Also list all pertinent data and information.

2. *Find:* State briefly what is to be determined.

3. *Diagram:* When applicable, draw a schematic, free body diagram (FBD), or sketch of the system of interest for the problem. Label any pertinent information on the diagram. This is a convenient method of including information that would be too difficult or lengthy to describe in words.

4. *Solution:* Show all the solution work and explain clearly all steps. List all the *assumptions* and *major equations* used in solving the problem. This section should be a logical procedure indicating the solution process from problem infancy to the final answer.

5. *Answer:* Clearly indicate the final answer to the problem, including appropriate units. Put a box around your final answer or double underline it as appropriate.

As stated previously, this problem presentation format is used by engineers to present the problem in its entirety to an audience. This could mean anything from a homework problem submitted to a professor to a vast, multimillion-dollar bridge project. Regardless of the scale, the presented problem is never the result of the initial run through the problem but is a polished product created after the problem has been completed.

EXAMPLE #1

Solve the following equation for x:

$$2x^2 + 3x - 7 = 0$$

Given: $2x^2 + 3x - 7 = 0$

Find: Solve for x.

Solution: Apply the quadratic formula

$$x = \frac{-b \pm \sqrt{b^2 - 4ac}}{2a}$$

where: $ax^2 + bx - c = 0$
therefore:

$$x = \frac{-3 \pm \sqrt{3^2 - 4 \times 2 \times (-7)}}{2 \times 2}$$

ANSWER: $x = 1.27, -2.77$

EXAMPLE #2

This example uses appropriate engineering problem format and introduces the concept of moments and static equilibrium. Static equilibrium requires that the sum of forces in any direction is equal to zero and that the sum of moments about any axis is also zero. For this example, we will use the sum of moments about the fulcrum of the teeter-totter equal to zero. In this two-dimensional problem, the moment due to a force is the product of the force and the perpendicular distance to the fulcrum.

Calculate the mass necessary to balance the beam shown.

Given: mass, dimensions (see figure 5.1)

Find: unknown mass

FBD: Teeter-totter

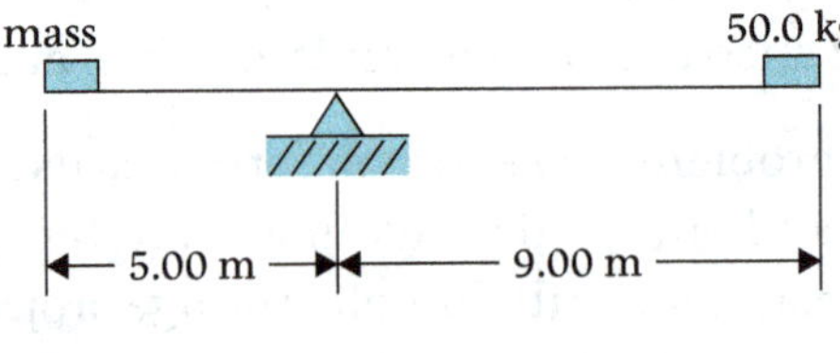

FIGURE 5.1

Solution: Static equilibrium, $\sum M_o = 0$

where M_o is the moment produced by each force about the pivot

Assumptions:

The mass of the beam is negligible.

Summing moments about the pivot:

$$\sum M_o = (\text{mass})g\,(5.00\text{ m}) - (50.0\text{ kg})(g)(9.00\text{ m}) = 0$$

where g = gravity = 9.81 m/s^2

$$\text{mass} = \frac{(50.0\text{ kg})(9.00\text{ m})}{(5.00\text{ m})} = 90.0\text{ kg}$$

ANSWER: mass = 90.0 kg

EXAMPLE #3

You are on an engineering team designing a water tank with a total volume of 6.00 x 105 L. The shape is to be cylindrical, with a hemispherical top. The cost to construct the cylindrical portion is $250/ m^2, while costs for the hemispherical portion are $350/m^2. What are the tank dimensions that will be the cheapest to build?

Given: total volume, $V_T = 6.00 \times 10^5$ L, hemisphere cost = $350/m^2, cylindrical cost = $250/m^2

Find: tank dimensions that will be cheapest to build

Diagram: Tank

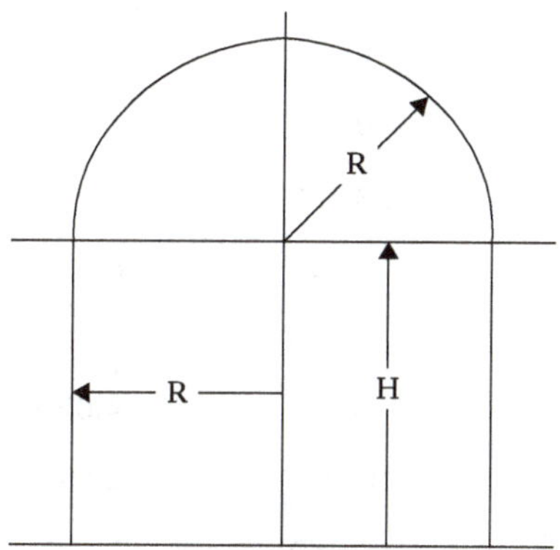

FIGURE 5.2 Cross-Section of Tank

Solution:

Volume of cylinder: $V_C = \pi R^2 H$

Volume of hemisphere: $V_H = \dfrac{2\pi R^3}{3}$

Surface area of cylinder: $SA_C = 2\pi RH$.

Surface area of hemisphere: $SA_H = 2\pi R^2$

Assumptions:

Tank is completely filled with water.

Construction costs and tank base cost are independent of size.

$$\text{Total volume} = V_T = V_C + V_H$$

$$6.00 \times 10^5 \text{ L} = 600 \text{ m}^3 = \pi R^2 H + \frac{2\pi R^3}{3}$$

$$\text{Note: } 1\text{m}^3 = 1000 \text{ L}$$

$$\text{Total cost} = C_T = C_C + C_H = \$250/\text{m}^2 \, (2\pi RH) + \$350/\text{m}^2 \, (2\pi R^2)$$

Combine the total cost equation and total volume equation to form one equation that yields cost as a function of radius:

$$C_T = \$250/\text{m}^2 \left[2\pi R \left(\frac{600 \text{ m}^2}{\pi R^2} - \frac{2R}{3} \right) \right] + \$350/\text{m}^2 (2\pi R^2)$$

Calculate cost for various radius sizes, or develop a table and graph:

Radius (m)	Cost ($)	Height (m)
1.00	301,152	190.32
2.00	154,608	46.41
3.00	110,367	19.22
3.50	99,825	13.26
4.00	93,431	9.27
4.50	89,993	6.43
5.00	88,798	4.31
5.07	88,782	4.05
5.50	89,391	2.65
6.00	91,469	1.31

From the table, the radius associated with minimum cost is approximately 5.07 m. From the equation for total volume, use the solved radius to find the associated height.

ANSWER: R = 5.07 m, H = 4.05 m

Note: this answer could be verified by using a derivative to minimize the cost equation:

$$\frac{dC}{dR} = 0, \text{ solve for R}$$

Summary

Engineers use a systematic approach to solving problems. First and foremost, the engineer must understand and formulate the problem. The process that engineers use supports this as well as the rest of the problem solving process. It is important, as illustrated in this chapter, to determine and clearly state the relevant information, what information is desired, assumptions made in formulating the problem or solution, and how the problem is solved. It is always prudent to check your work, and iterate if necessary.

Assignment

For the problems in this assignment, use the engineering problem solving format outlined above and demonstrated in the examples. Be sure all applicable steps are included. Clearly state any assumptions. Use appropriate conversions and number of significant figures.

1. In designing a radio tower at an airport, engineers must take into account the ascent rate (slope at which a plane can take off) such that planes will not collide with the tower. If the most shallow ascent rate is found to be 9 percent (the plane climbs vertically 9 ft for every 100 ft of horizontal travel) and the tower is 300 ft tall, how far can the tower be safely placed from the end of the runway?

2. Katie just bought new tires for her car. If the diameter of the tires is 2 ft and the width is 8 in, how many revolutions will her tires make if she drives 1.5 miles (1 mi = 5280 ft)?

3. Use the diagram below to answer the question that follows.

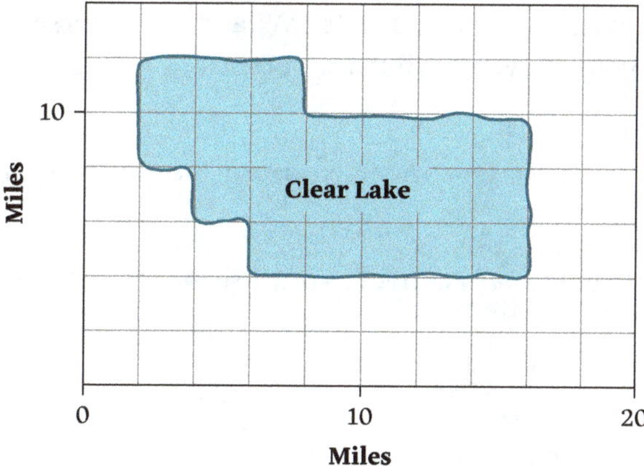

What is the total length of Clear Lake's shoreline?

4. Use the diagram below to answer the question that follows.

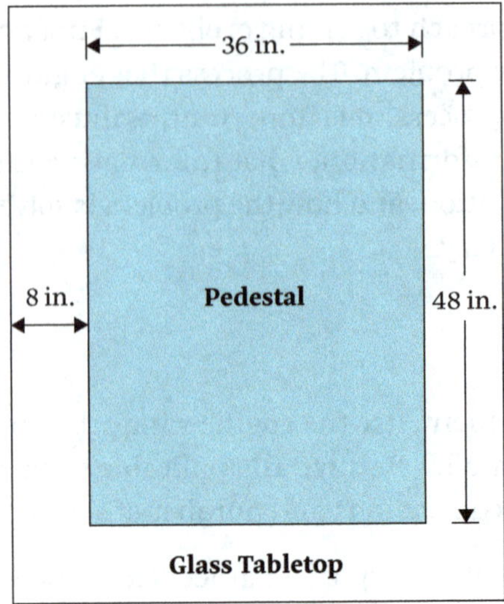

A glass tabletop is supported by a rectangular pedestal. If the tabletop is 8 inches wider than the pedestal on each side, what is the perimeter of the glass tabletop?

5. Peter works 38 hours per week and earns $7.25 per hour. His employer gives him a raise that increases his weekly gross pay to $307.80. What is the increase (in dollars and percentage) in Peter's weekly gross pay? Also, what is his new hourly rate?

6. The Lincoln Reflecting Pool in Washington, DC, holds approximately 6.75×10^6 gallons of water. Calculate the a) weight (lb) and b) volume (ft^3) of water in the reflecting pool. Note that water has a unit weight (γ) of 62.4 lb/ft^3. What other information do you need? How confident are you in your answer? What are potential sources for error in your method of calculation?

7. Using the results of Problem #6, convert your answer for weight into terms of a) kips and b) kN.

8. Using the results of Problem #6, convert your answer for volume into terms of a) liters and b) m^3.

Handling Technical Information

Objectives

- Determine the process to collect and record technical data.
- Organize and perform calculations on technical data using a spreadsheet utility (i.e., Microsoft Excel). Note: This module is based on the Windows 2016/Office 365 versions of Excel. For Mac, Android, or iOS versions of Excel, consult the program's Help menu.
- Utilize a spreadsheet utility to create professional-looking graphs for interpretation of data.
- Determine data relationships by use of the *trendline* function.

When engineers solve problems, inside or outside the classroom, they must understand how to handle the information or data relevant to that problem. The ability to collect, record, and represent that data in an organized and meaningful way is of utmost importance.

Collecting and Recording Data by Hand

The most basic level of handling data and information is done by hand. Even in today's world of pocket computers, in the laboratory, most engineers carry notebooks. During an experiment, a pencil and paper are an engineer's most basic tools for recording not only results but all test conditions and details. Below is a simple example to introduce you to data collection.

EXAMPLE #1

The information below was collected by students in a class experiment in which they launched projectiles. Could you recreate this experiment? What is missing?

Distance (ft)	Time (s)
90	3.5
120	6
75	2.5
80	3
135	9

It is important to know that an experiment yields more than just results. The students did a good job of recording data, experimental conditions (distance, time), and units (feet, seconds), but all other aspects of the test are unknown. What was launched? How was it launched? At what angle? Does anything else need to be considered? Much more detail is needed to recreate this experiment.

The next step in handling experimental data is to represent it in a form in which it can be most easily and effectively understood. This is most often accomplished using graphs. The simplest graphs are done freehand as a first representation of data. Here are key points to remember when plotting data by hand:

- Label axes for scale, experimental condition, and units.
- Try to center the data in the plot as reasonably as possible.
- Be sure axis scale is consistent.
- Use a straightedge when drawing any lines.
- Organization and legibility are essential.

EXAMPLE #2

Use the data from the previous example problem to construct a freehand graph of the results of the test.

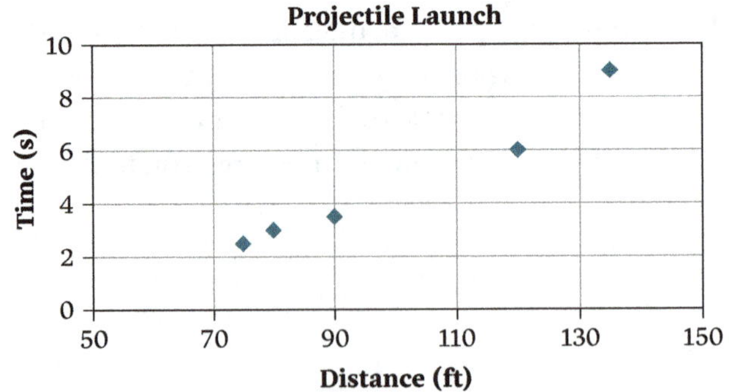

Does seeing the test results in a graph tell you more than just the table did? Do you notice any trends in the data?

Microsoft Excel

Microsoft Excel is a computational spreadsheet program. Excel is extremely useful for the following functions:

- Recording and presenting data in a very orderly format
- Creating graphs of data
- Performing the same calculations repeatedly
- Data analysis (trend fits and numerical solutions)
- Working with tables of information (material properties)

Excel does not do symbolic mathematical manipulation, and though it is capable of evaluating logical statements, it is not designed as a programming language. Other programs, such as Matlab, MathCAD, and Mathematica, are better suited for these tasks.

After recording and plotting data by hand, many engineers will use Excel as a more powerful, permanent, and organized method of handling, analyzing, and presenting information. As an introduction to Excel, let's continue the previous example.

EXAMPLE #3

Enter the previously collected data (from Example #1) into an Excel spreadsheet.

To enter text into a cell in the sheet, simply click on the cell and type. (Enter "Distance" in cell B3, etc.). Text can also be entered in the *formula bar.* Keyboard arrows or the tab and return keys can also be used to move cells. Entered into Excel, the data should look like this:

Distance (ft)	Time (s)
90	3.5
120	6
75	2.5
80	3
135	9

Here is a screen shot of the entered data:

	Distance (ft)	Time (s)
	90	3.5
	120	6
	75	2.5
	80	3
	135	9

One of the strengths of Excel is in making repeated calculations. Using an equals sign (=) as the lead character in a cell indicates to Excel that you are entering a formula and you would like it to perform a calculation. For example, entering "=2 + 3" into a cell would cause the cell to then display the number 5, which is the answer to the formula of 2 + 3. Calculations can also be made using a cell's address to identify the number (in the data entered above, the cell address B5 identifies a value of 90).

EXAMPLE #4

Use Excel to calculate the average velocity of the first projectile (see Example #3). Hint: velocity = distance/ time.

To enter the formula, click cell D5 and type "=B5/C5," then hit "Enter". Note that the result of the formula is displayed in cell D5, but the formula is displayed in the *formula bar*. Now try an alternate method; click cell D5, type "=," then click cell B5. What happened?

	A	B	C	D	E
1					
2					
3		Distance (ft)	Time (s)	Velocity (ft/s)	
4					
5		90	3.5	25.71429	
6		120	6		
7		75	2.5		
8		80	3		
9		135	9		

Notice the small square at the bottom right corner of the active cell. This is called the *fill handle*, and it serves as an essential shortcut to copying cells. Click on the square and drag your cursor and it will copy the contents of the active cell into all highlighted cells. If the active cell contains a formula, it will increment the formula in the direction of the copy. This can be useful when performing the same calculation on large quantities of data.

EXAMPLE #5

Use Excel to calculate the average velocity of all projectiles (see Example #3).

Click on the fill handle of cell D5 and drag the cursor through cell D9.

D6					f_x	=B6/C6	
	A	B	C	D	E		
1							
2							
3		Distance	Time		Velocity		
4		(ft)	(s)		(ft/s)		
5		90	3.5	25.71429			
6		120	6	20			
7		75	2.5	30			
8		80	3	26.66667			
9		135	9	15			
10							
11							

Everything is technically correct in the spreadsheet above; however, there is a problem with the number of significant digits displayed in the results in column D.

EXAMPLE #6

Display the velocities with one decimal place.

To change how Excel displays a cell, select a cell (or cells), right-click, and select "Format Cells." Select the "Number" category, and enter "1" as the decimal place. What other formatting changes can be made in this window?

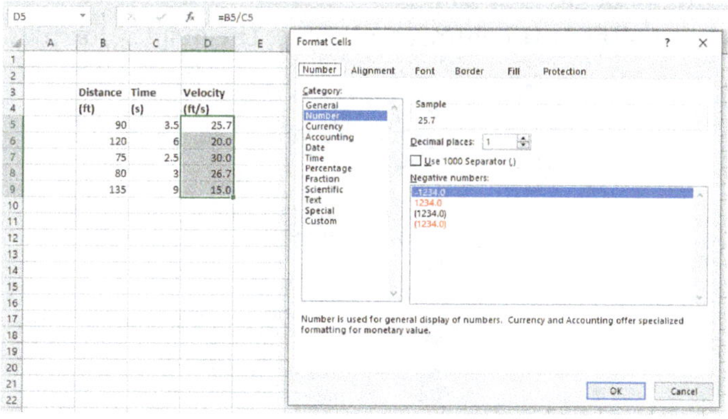

Excel has the ability to recognize both relative and absolute addressing. This means that when copying a formula, it can increment some addresses while holding others constant, as specified by the user. A dollar sign preceding a row and/or column address indicates to Excel that that address is to be held constant.

EXAMPLE #7

Display the projectile velocities in units of in/s.

Enter a conversion factor of 12 in/ft in cell B1. In cell E5, type "=D5*B1." Use the fill handle to copy this formula to cells E6–E9 (results are shown below). Now learn an alternate method: in cell E5, type "=," click cell D5, type "*," click cell B1, and then press the F4 key. What happened? Press F4 again. Then press F4 again. What happens?

	A	B	C	D	E	F
1	Conv Fact		12 in/ft			
2						
3		Distance	Time		Velocity	Velocity
4		(ft)	(s)		(ft/s)	(in/s)
5			90	3.5	25.7	308.5714
6			120	6	20.0	240
7			75	2.5	30.0	360
8			80	3	26.7	320
9			135	9	15.0	180
10						

Cell E9: =D9*B1

Excel also has the capability of letting you assign variable names to values. The script in the Excel *Name Box* can be changed such that a variable name may be used in formulas rather than the cell address. Named cells are always treated as absolute addresses by Excel.

EXAMPLE #8

Use a variable name for the unit conversion factor.

Click cell B1. Type "ConvFactor" in the *Name Box*. Alter the formula in cell E5 to "=D5*ConvFactor." Copy this formula for the rest of the column and adjust the display format.

	A	B	C	D	E	F
1	Conv Fact		12 in/ft			
2						
3		Distance	Time		Velocity	Velocity
4		(ft)	(s)		(ft/s)	(in/s)
5			90	3.5	25.7	308.6
6			120	6	20.0	240.0
7			75	2.5	30.0	360.0
8			80	3	26.7	320.0
9			135	9	15.0	180.0
10						

Name Box: ConvFact, value: 12

Excel comes equipped with a large number of useful built-in functions. These functions can be typed into the *formula bar* or chosen from the *Formulas* tab. The object of a function is designated by parentheses. If a range of cells or a scattered selection of cells are the object, they are seperated by a colon or a comma, respectively. These objects can be typed into the formula or selected using the mouse.

EXAMPLE #9

Calculate the average velocity and standard deviation of the projectiles.

In cell D11, type "=AVERAGE(D5:D9)," or select the *Formulas* tab, select *Average*, and use the mouse to select the desired range of cells (there are actually several different ways to accomplish this task). This formula can then be copied to cell E11. The procedure is the same for standard deviation. What other Excel functions might you find useful?

D12	▼ ⋮	✕ ✓	f_x	=STDEV(D5:D9)

	A	B	C	D	E	F
1	Conv Fact		12 in/ft			
2						
3		Distance	Time	Velocity	Velocity	
4		(ft)	(s)	(ft/s)	(in/s)	
5		90	3.5	25.7	308.6	
6		120	6	20.0	240.0	
7		75	2.5	30.0	360.0	
8		80	3	26.7	320.0	
9		135	9	15.0	180.0	
10						
11			Avg	23.5	281.7	
12			St Dev	6.0	71.4	
13						

Graphing with Excel

One of the most useful and widely used aspects of Excel is its graphing utility. It allows users to create a wide variety of graphs, with wide control over options of the display of the information. These features are essential when presenting data and information to clients, customers, or employers in the engineering world.

Many of the same principles that applied to freehand graphs still apply in Excel:

- Label axes for scale, experimental condition, and units.
- Try to center the data in the plot as reasonably as possible.
- Organization and clarity are essential.

The most commonly used graph in Excel for engineering purposes is a scatter plot. In its simplest form, this is an X–Y plot that requires two columns or rows of data of the same size.

	A	B	C	D
1	Component Heating			
2				
3	Time (s)	Temp. (K)		
4	0	300		
5	5	498		
6	10	633		
7	15	761		
8	20	902		
9	25	998		
10	30	1090		
11	35	1178		
12	40	1251		
13	45	1309		
14	50	1364		
15	55	1400		
16	60	1439		
17	65	1463		
18	70	1489		
19	75	1509		
20	80	1511		
21	85	1528		
22	90	1532		
23	95	1544		
24	100	1550		
25				

EXAMPLE #10

Copy the data above into an Excel spreadsheet and create a scatter plot of the data.

Using the mouse, select the range of cells from A4 to B24. Select the *Insert* tab, click *Scatter*, then *Scatter with only Markers*.

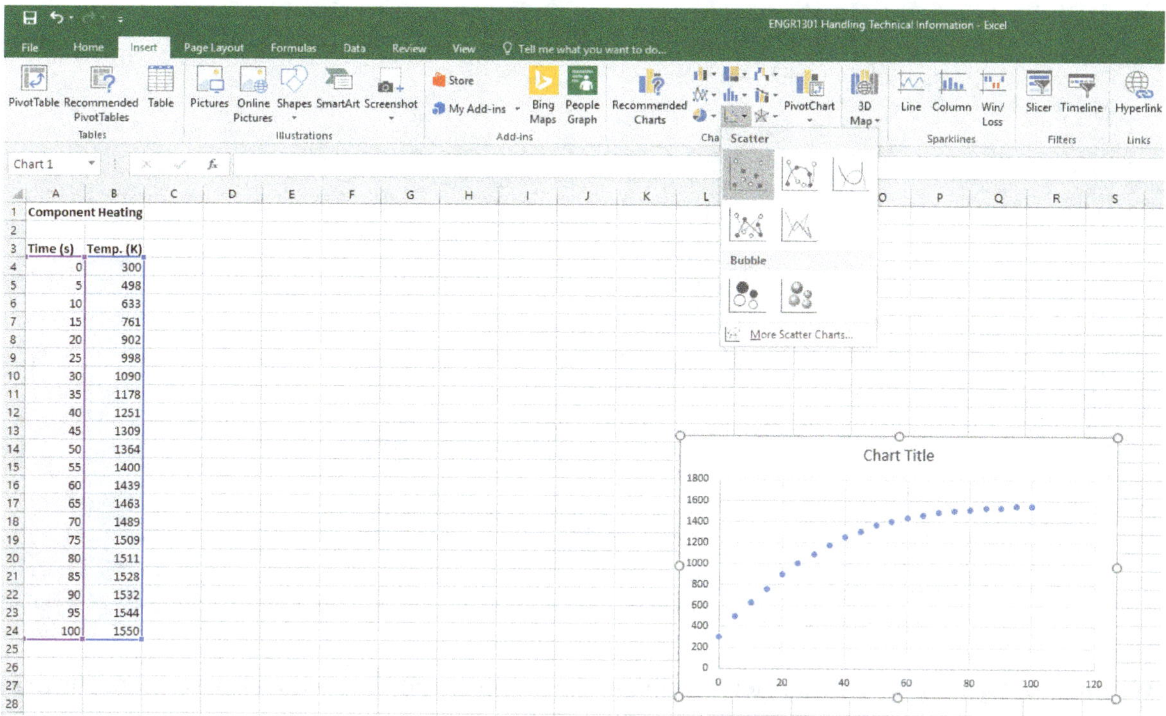

The result is shown below (see Fig. 6.10). Now, this is a plot of the data, but it fails miserably at conveying the results of the experiment. There is no context, and without context, the data is useless. Right-click on the chart to Move the chart to its own worksheet and click OK. This will allow for a full-sized graph for printout. Otherwise, the chart will remain on the current worksheet.

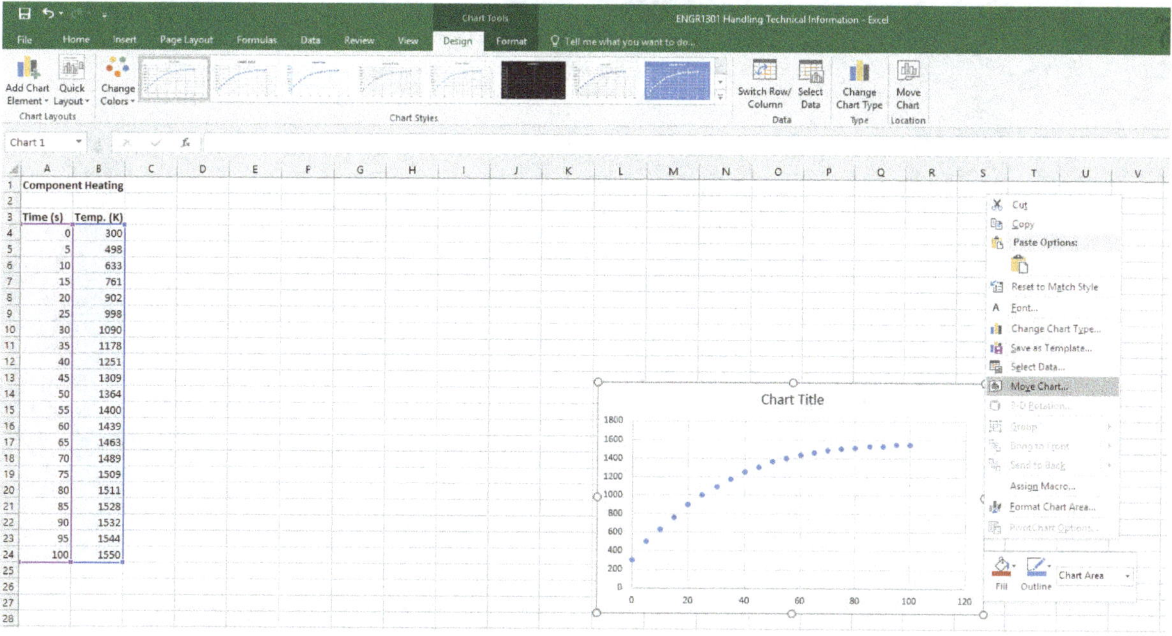

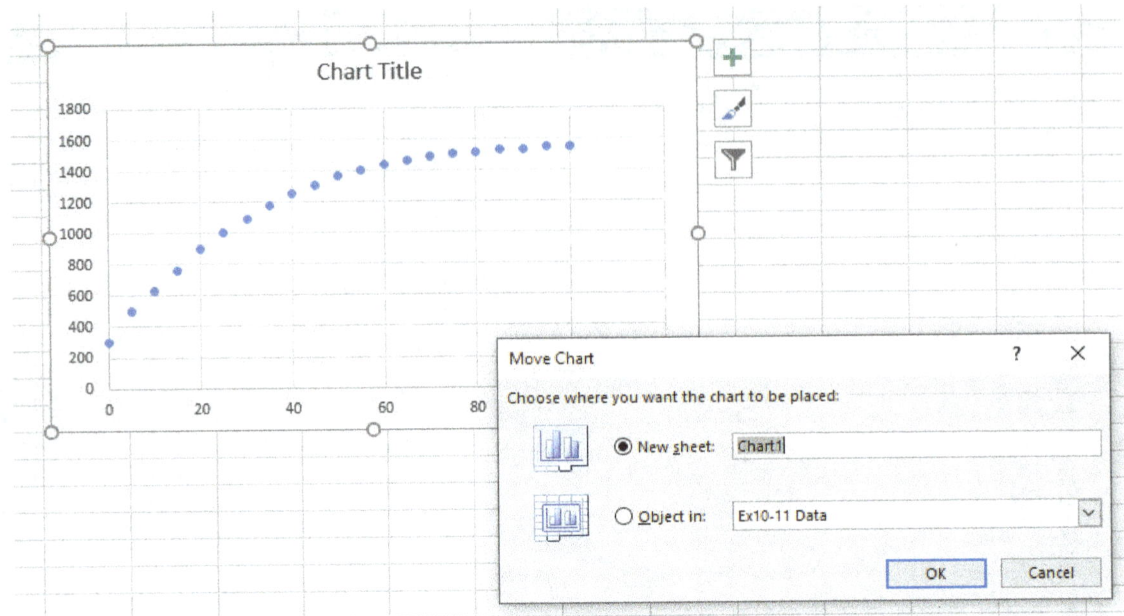

We must add a chart and axes titles, legend, and gridlines, and then adjust the range of the axes. Note: If there is a box called Series 1 to the right of the graph, you may click on and delete it.

To insert a chart title, select the Design tab, click Add Chart Element, then Chart Title. Use the Above Chart option on the right side. Double click on the title text box to enter the title "Component Heating."

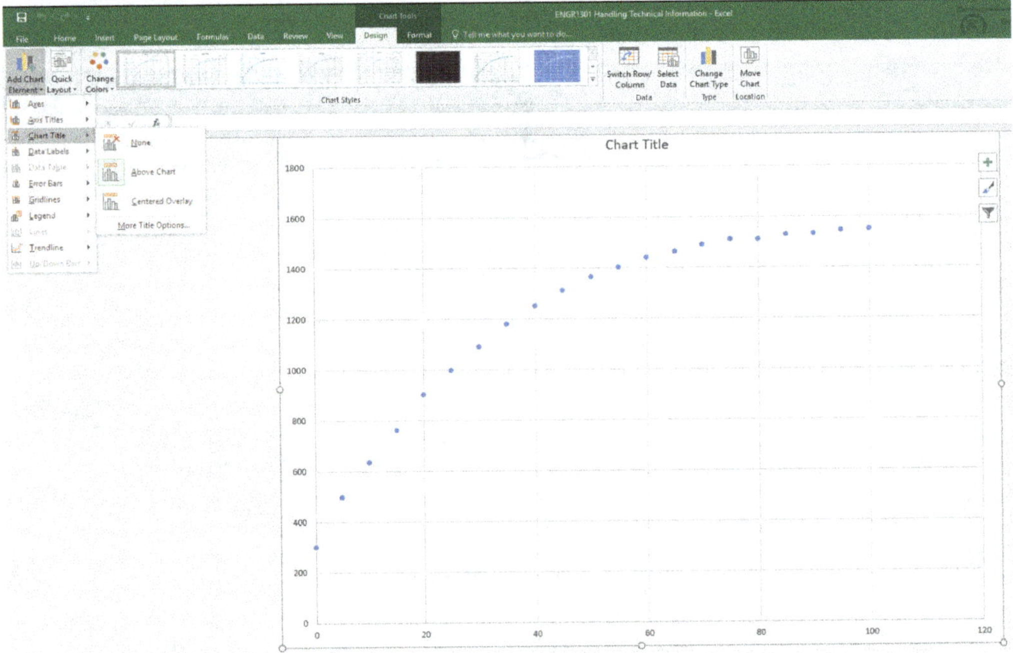

To enter axes labels, click the Design tab, click Add Axis Title, and choose the Primary Vertical axis option, then enter a label, "Temp (K)," in the text box beside the axis. Repeat process for the horizontal axis, "Time (s)".

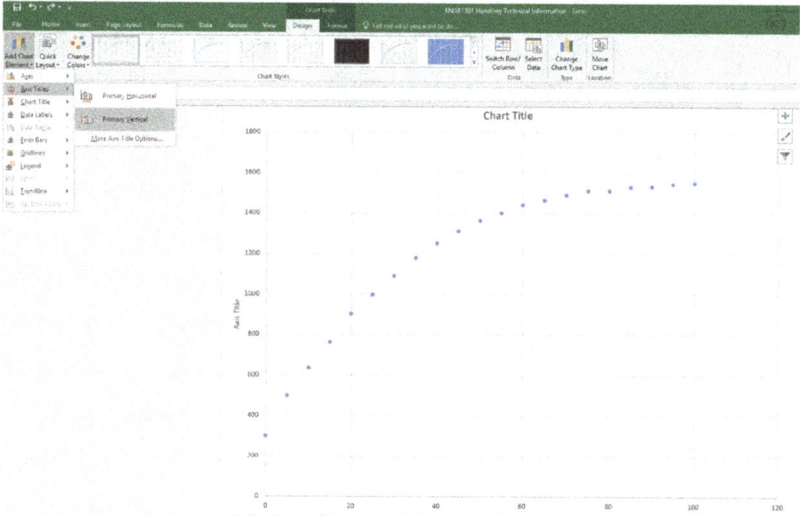

To toggle the gridlines, click the Design tab, click Gridlines, and choose Primary Major Vertical Gridlines. If they were displayed, they will turn off. Toggle again to make sure that they are on. The range of the axes needs to be adjusted to eliminate vacancies in the chart, center the data, and thus make the plot more visually appealling. Right-click on the X-axis and select Format Axis. Under Axis Options, enter a maximum fixed value of 100 and a minimum of 0. Change the minor axis value to 2.0, if it isn't already updated. Toggle the Tick Marks for Major (Outside) and Minor (Inside). Repeat this process for the Y-axis and enter a range of 0–1600. Set the Y-axis minor value to 50.

To toggle on the minor axes, right-click on each axis and select Add Minor Gridlines. (X-axis modifications are shown below.)

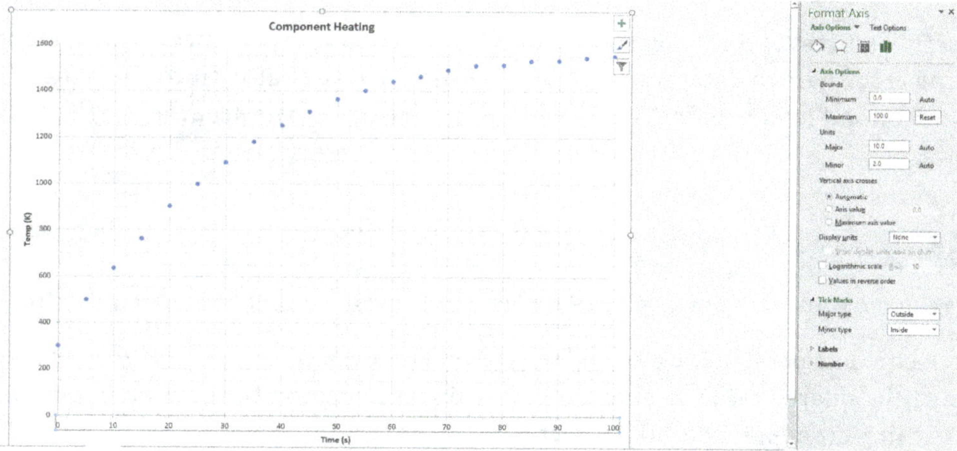

The gridlines and axes may be too faint to see. To fix this, right click on each axis and select the Format Axis menu, then click on the icon on the left that looks like a paint can. Scroll to Line and then adjust the Width from the default value. A good value for the axes themselves is to increase from the default of 0.75 pt to 2 pt. Repeat for the gridlines by right-clicking on them, and adjust in the same manner. A good value for the major gridlines is to increase from 0.75 pt to 1 pt. The final graph is shown below.

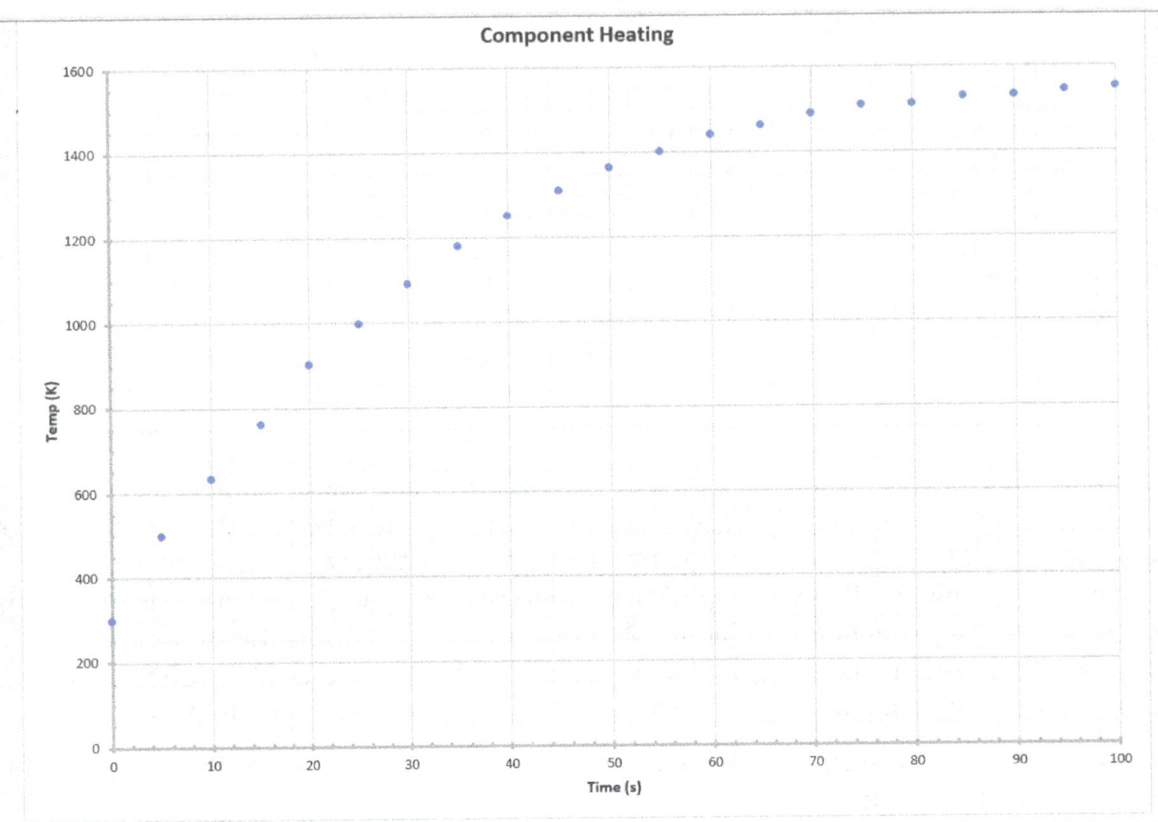

Other useful features of Excel's graphing utility are plotting trendlines, evaluating R^2 values, and adding error bars to data. A *trendline* is a line or curve that is fit to the data to help predict or understand data trends. The R^2 value is a measure of the accuracy of the trendline fit to the data.

EXAMPLE #11

Insert a trendline with an R^2 value > 0.999 and error bars in the graph from Example #10.

Right-click on a data point in the plot and select Add Trendline. Select "Display Equation on chart" and "Display R-squared value on chart." Neither linear nor second-order polynomial yield an R^2 > 0.999, so third-order polynomial is selected.

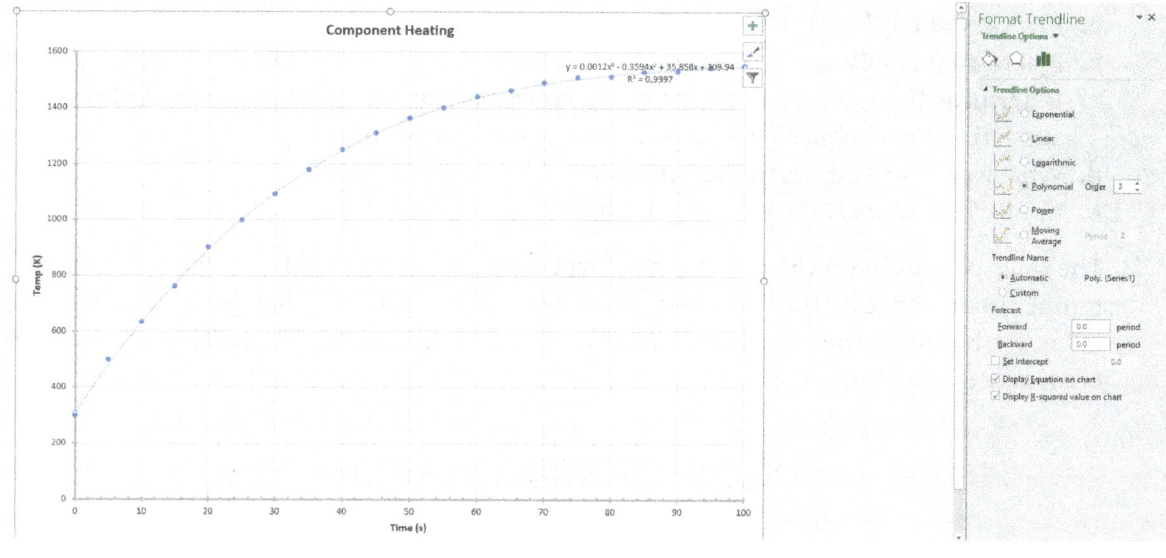

Summary

Engineers must process technical data for calculations used in the analysis and design of engineered systems. A spreadsheet utility such as Microsoft Excel can be used to tabulate data, perform calculations, and present the results in a professional graphical format. Trends such as linearity (i.e., $y = mx + b$), power series, and others can be discerned by the graphical interpretation of the data. The resulting trendline can be used to predict results for similar data.

Assignment

1. The table below shows the results of a test of an experimental aircraft.

Time (s)	Velocity (m/s)
10	20.2
20	42.1
30	84.4
40	112.4
50	157.3
60	185.3

   a. Plot the data by hand.

   b. Plot the data in Excel.

   c. Determine the equation of the line. What does the nature of the data curve tell you about the experiment? What does the slope of the line represent for this test?

   d. What would you expect the velocity of the vehicle to be at 80 s?

   e. Use Excel to convert the velocities to mph.

2. The table below shows the results of an experiment using a thermocouple to measure temperature in a test fire. The thermocouple is calibrated such that the temperature is known based on the voltage difference in the wires. The wire manufacturer gives an error rating of ±3%.

Voltage (mV)	Temp (K)
2	41
4	64
6	97
8	139
10	177
12	212
14	253
16	286
18	331
20	368

   a. Plot the data by hand.

   b. Plot the data in Excel.

   c. Determine the equation of the line. What does the nature of the data curve tell you about the experiment?

   d. What voltage would you expect to see if you knew the fire was burning at 400K? How certain are you in your answer? Is there a way to quantify this certainty?

   e. Use Excel to convert the temperatures to Fahrenheit.

3. During a wind tunnel test in a fluid mechanics laboratory, the students measured the drag force on a golf ball. The resulting data is shown below.

Velocity (ft/s)	Force (lb$_f$)
5	0.001
10	0.0024
15	0.0042
20	0.0065
25	0.0092
30	0.012
35	0.015
40	0.019
45	0.024
50	0.028
55	0.034
60	0.039

a. Plot the data in Excel.

b. What does the nature of the data curve tell you about the experiment? What type of trendline best fits the data?

c. An object's terminal velocity is the maximum and constant speed an object reaches due to the restraining force exerted by the air, water, or other fluid through which it is moving. In this case, the terminal velocity would occur when the weight of the golf ball equals the drag force (W = m*g). At what velocity would you expect this to occur for the golf ball? How certain are you in your answer? Is there a way to quantify this certainty?

d. Use Excel to convert the drag force to Newtons.

4. To determine the stress in an engineering material, you only need to know the force on that object and its dimensions.

$$\sigma = F/A$$

where

σ = axial stress

F = force

A = cross-sectional area

You are on an engineering team designing an elevator system, and you have been tasked to pick the cable to be used. You decide to use a steel cable (σ = 165 MPa), but the client is uncertain of the weight rating they want for their elevator.

a. Use Excel to create a table of data that shows the cable diameter needed for elevator weights ranging from 100 to 300 kN in increments of 10 kN.

b. If the largest cable that the pulley system can accommodate is a 4 cm cable, what is most that the elevator can weigh?

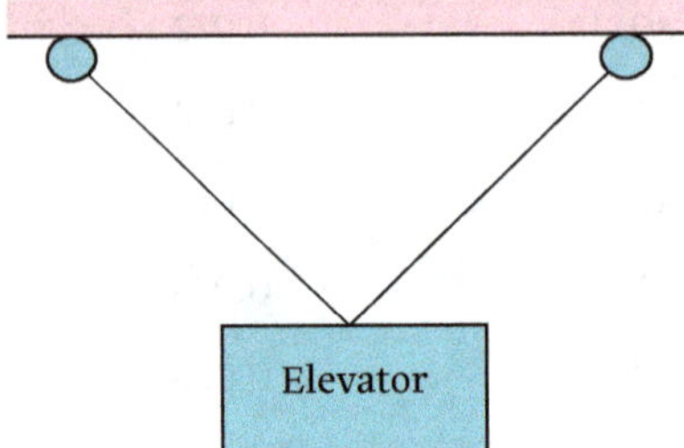

5. A 50 kg block is suspended from rings by two identical cables, as shown in Figure 6.17. For this system, the tension in the cable is calculated using the following equation:

$$T = \frac{mg}{2 \sin \theta}$$

where

T = tension

m = mass of the block

g = gravity (9.81 m/s^2)

a. If the distance between the rings is changed to vary the value of θ between 20° and 80°, use Excel to create a table of data that shows the cable tensions for this range of angles. (Hint: use Excel's built-in trig functions, but be careful, because Excel uses *radians* as the standard unit for angles.)

b. If the maximum tension the cables can hold is 650 N and the length of the cables is 1.2 m, what is the maximum distance the rings can be spaced without the cables breaking?

Statistics

Objectives

- Identify the *normal distribution function (bell curve).*
- Define the following terms in regard to the normal distribution function:
  - *Mean* (average)
  - *Mode*
  - *Variance*
  - *Standard deviation*
- Utilize statistical functions to analyze data that follow a normal distribution model.
- Describe the *coefficient of determination* (R^2).

Normal Distribution Function

Engineers gather *data* on a continual basis as a part of their jobs. These data (*data* is a plural word) can be characterized as a *population* that is inclusive of all pieces of data. For example, all of the 575 males in a school would be a population of known size. A *sample* is a smaller set of data points taken such that the sample is representative of the total population. For example, a sample could be 50 males selected randomly from the school population of 575 males.

The *normal distribution function* (or *bell curve*), as shown in Figure 7.1 below, describes a large population data set, as well as randomly chosen sample data sets. Its bell-curved shape represents the tendency for data to group about a central value (the *mean* or *average*), with fewer and fewer data points away from the central value.

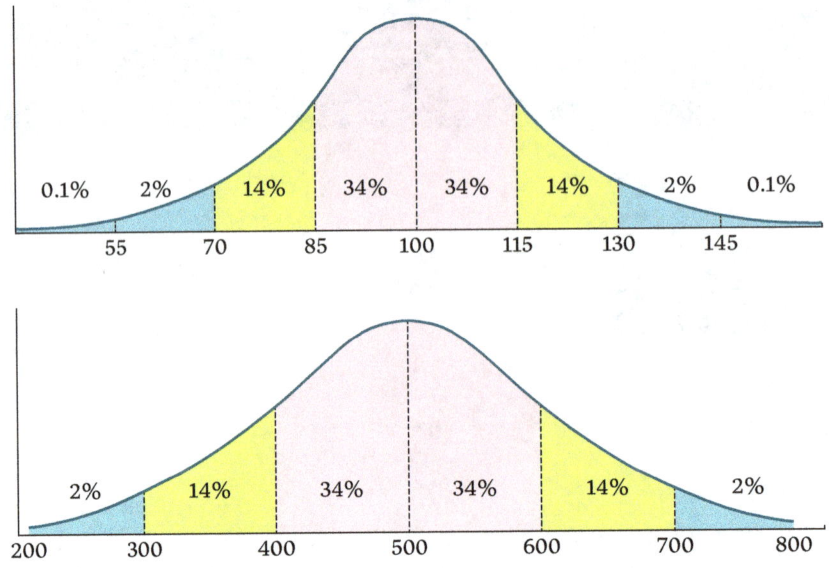

FIGURE 7.1 The Normal Distribution Function.

Two types of statistics can be derived from the normal distribution function. *Descriptive statistics* can be used to describe characteristics of a sample (its mean value, its standard deviation, etc.) *Inferential statistics* uses information about the sample to make educated guesses (inferences) about future samples.

The three most important descriptive statistical calculations are the mean, median, and standard deviation. The *mean* (*average*) and *median* are both measures of the location or central tendency of the sample. *Standard deviation* is a measure of the variability or spread of the sample data. The square of the standard deviation is called the *variance*. These are shown by the use of examples in the following discussion.

Mean (Average)

The *mean* (*average*) of a sample can be calculated by adding up all the values of the sample and dividing the sum by the number of elements in the sample.

$$\mu = \frac{x_1 + x_2 + \ldots + x_n}{n}$$

The *location* of a random sample is related to the accuracy of a measurement. For example, if you are measuring the average weight of males in a school population, do not just weigh males who are on the football team, unless you are specifically targeting this subset of the population. Such data could be *biased* or skewed in relation to the total population.

Median

The *median* is the middle value in a set of data that is arranged in ascending order. For an even number of data points, the median is the average of the middle two data points.

Standard Deviation

The *standard deviation* is a measure of the variability of the data. The *variability* of a random sample is related to the precision of a measurement. The *sample standard deviation*, *s*, of *n* values can be readily calculated. The *population standard deviation*, σ, is a better estimate of the variability, as it considers every observation in a data set. The *variance* is the square of *s* or σ, depending on which is needed.

$$s = \sqrt{\frac{\sum_{i=1}^{n}(x_i - \bar{x})^2}{n-1}} \qquad \sigma = \sqrt{\frac{\sum_{i=1}^{n}(x_i - \bar{x})^2}{n}}$$

Note: The symbol *s* is used for the standard deviation of a sample. The symbol σ is used for the standard deviation of a population (with division by *n* instead of *n* – 1), especially when using the true population mean μ. As the number of data points increases, the mean approaches μ, and the standard deviation of a sample approaches σ. Practically, many textbooks use the symbol σ when discussing standard deviation.

In Microsoft Excel, use the STDEV function for *s* and STDEV.P function for σ. Excel uses the AVERAGE function to find the mean and the MEDIAN function to find the median value.

EXAMPLE #1

Given: the following data sample: X = {–2, 3, 1, 4, –1, 10}

Find: the mean, median, and standard deviation of this sample

Solution:

Mean:

$$\bar{x} = \frac{-2+3+1+4-1+10}{6} = 2.5$$

Median:

Sorted Data: –2, –1, 1, 3, 4, 10

Median = (1 + 3)/2 = 2

Sample Standard Deviation:

$$s = \sqrt{\frac{\sum_{i=1}^{n}(x_i - \overline{x})^2}{n-1}}$$

$$s = \sqrt{\frac{(-2-2.5)^2 + (3-2.5)^2 + (1-2.5)^2 + (4-2.5)^2 + (-1-2.5)^2 + (10-2.5)^2}{5}}$$

$$s = \sqrt{\frac{(-4.5)^2 + (0.5)^2 + (-1.5)^2 + (1.5)^2 + (-3.5)^2 + (7.5)^2}{5}}$$

$$s = \sqrt{\frac{20.25 + 0.25 + 2.25 + 2.25 + 12.25 + 56.25}{5}}$$

$$s = \sqrt{\frac{93.5}{5}} = 4.32$$

<u>Population Standard Deviation</u>

$$\sigma = \sqrt{\frac{93.5}{6}} = 3.94$$

Inferential Statistics

Inferential statistics can be used to solve a variety of problems:

- What is the likelihood that a resistor from this bag of 100 Ω resistors exceeds 120 Ω?
- If this concrete cylinder was supposed to break at 100 kips (100,000 lb) but actually broke at 75 kips (75,000 lb), was that just coincidence, or is there something wrong with this batch of concrete?
- If this machine has a mean time between failures of 1,000 hours, what is the chance it will break down within 500 hours?
- If my bowling average is 180 and I'm bowling against someone whose average is 190, what is the probability that I will win?

Probability

The *probability* of an unknown event happening is its likelihood. This is assigned a number between 0 and 1. The value of 0 corresponds to a 0% chance, while the value of 1 corresponds to a 100% chance. If there is a 10% chance that event A will happen, we say that P(A) = 0.1. The sum of the probabilities of all possible outcomes is always 1. If A, B, C, and D are the only four possible outcomes, then P(A) + P(B) + P(C) + P(D) = 1. This topic will be expanded on in the next section of this textbook.

Use of the Normal Distribution

Normal distributions are useful whenever there are a large number of *independent* factors that can contribute to the outcome. Probabilities for normal distributions are typically looked up in a z-table, shown in Table 7.1 and 7.2. The table gives values for the standard normal curve, which has a mean of 0 and a standard deviation of 1. The values in the table can be scaled for any other data set with a known mean and standard deviation value.

A normal distribution usually refers to a large, continuous population. The *y*-axis is referred to as a *density function*, or the probability that an event will occur at that *x*-value. The probability is an area in the diagram, as shown below. For example, the probability that someone in the population is between 5 ft 2 in and 5 ft 8 in in height would be an area in the diagram bounded by two lines. Note that the mean and standard variation for a sample are $\bar{x}$ and *s*, *respectively*, and the mean and standard variation of a population are μ and σ, respectively.

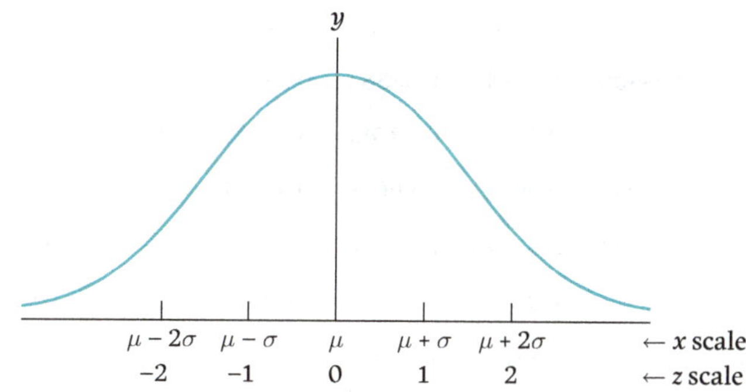

FIGURE 7.2 Normal Distribution in Terms of Mean and Standard Deviations.

TABLE 7.1 Positive Z-values (normalized area to the *left*, where total area = 1)

z	0	0.01	0.02	0.03	0.04	0.05	0.06	0.07	0.08	0.09
0.0	0.50000	0.50399	0.50798	0.51197	0.51595	0.51994	0.52392	0.52790	0.53188	0.53586
0.1	0.53983	0.54380	0.54776	0.55172	0.55567	0.55962	0.56356	0.56749	0.57142	0.57535
0.2	0.57926	0.58317	0.58706	0.59095	0.59483	0.59871	0.60257	0.60642	0.61026	0.61409
0.3	0.61791	0.62172	0.62552	0.62930	0.63307	0.63683	0.64058	0.64431	0.64803	0.65173
0.4	0.65542	0.65910	0.66276	0.66640	0.67003	0.67364	0.67724	0.68082	0.68439	0.68793
0.5	0.69146	0.69497	0.69847	0.70194	0.70540	0.70884	0.71226	0.71566	0.71904	0.72240
0.6	0.72575	0.72907	0.73237	0.73565	0.73891	0.74215	0.74537	0.74857	0.75175	0.75490
0.7	0.75804	0.76115	0.76424	0.76730	0.77035	0.77337	0.77637	0.77935	0.78230	0.78524
0.8	0.78814	0.79103	0.79389	0.79673	0.79955	0.80234	0.80511	0.80785	0.81057	0.81327

0.9	0.81594	0.81859	0.82121	0.82381	0.82639	0.82894	0.83147	0.83398	0.83646	0.83891
1.0	0.84134	0.84375	0.84614	0.84849	0.85083	0.85314	0.85543	0.85769	0.85993	0.86214
1.1	0.86433	0.86650	0.86864	0.87076	0.87286	0.87493	0.87698	0.87900	0.88100	0.88298
1.2	0.88493	0.88686	0.88877	0.89065	0.89251	0.89435	0.89617	0.89796	0.89973	0.90147
1.3	0.90320	0.90490	0.90658	0.90824	0.90988	0.91149	0.91308	0.91466	0.91621	0.91774
1.4	0.91924	0.92073	0.92220	0.92364	0.92507	0.92647	0.92785	0.92922	0.93056	0.93189
1.5	0.93319	0.93448	0.93574	0.93699	0.93822	0.93943	0.94062	0.94179	0.94295	0.94408
1.6	0.94520	0.94630	0.94738	0.94845	0.94950	0.95053	0.95154	0.95254	0.95352	0.95449
1.7	0.95543	0.95637	0.95728	0.95818	0.95907	0.95994	0.96080	0.96164	0.96246	0.96327
1.8	0.96407	0.96485	0.96562	0.96638	0.96712	0.96784	0.96856	0.96926	0.96995	0.97062
1.9	0.97128	0.97193	0.97257	0.97320	0.97381	0.97441	0.97500	0.97558	0.97615	0.97670
2.0	0.97725	0.97778	0.97831	0.97882	0.97932	0.97982	0.98030	0.98077	0.98124	0.98169
2.1	0.98214	0.98257	0.98300	0.98341	0.98382	0.98422	0.98461	0.98500	0.98537	0.98574
2.2	0.98610	0.98645	0.98679	0.98713	0.98745	0.98778	0.98809	0.98840	0.98870	0.98899
2.3	0.98928	0.98956	0.98983	0.99010	0.99036	0.99061	0.99086	0.99111	0.99134	0.99158
2.4	0.99180	0.99202	0.99224	0.99245	0.99266	0.99286	0.99305	0.99324	0.99343	0.99361
2.5	0.99379	0.99396	0.99413	0.99430	0.99446	0.99461	0.99477	0.99492	0.99506	0.99520
2.6	0.99534	0.99547	0.99560	0.99573	0.99585	0.99598	0.99609	0.99621	0.99632	0.99643
2.7	0.99653	0.99664	0.99674	0.99683	0.99693	0.99702	0.99711	0.99720	0.99728	0.99736
2.8	0.99744	0.99752	0.99760	0.99767	0.99774	0.99781	0.99788	0.99795	0.99801	0.99807
2.9	0.99813	0.99819	0.99825	0.99831	0.99836	0.99841	0.99846	0.99851	0.99856	0.99861
3.0	0.99865	0.99869	0.99874	0.99878	0.99882	0.99886	0.99889	0.99893	0.99896	0.99900
3.1	0.99903	0.99906	0.99910	0.99913	0.99916	0.99918	0.99921	0.99924	0.99926	0.99929
3.2	0.99931	0.99934	0.99936	0.99938	0.99940	0.99942	0.99944	0.99946	0.99948	0.99950
3.3	0.99952	0.99953	0.99955	0.99957	0.99958	0.99960	0.99961	0.99962	0.99964	0.99965
3.4	0.99966	0.99968	0.99969	0.99970	0.99971	0.99972	0.99973	0.99974	0.99975	0.99976
3.5	0.99977	0.99978	0.99978	0.99979	0.99980	0.99981	0.99981	0.99982	0.99983	0.99983
3.6	0.99984	0.99985	0.99985	0.99986	0.99986	0.99987	0.99987	0.99988	0.99988	0.99989
3.7	0.99989	0.99990	0.99990	0.99990	0.99991	0.99991	0.99992	0.99992	0.99992	0.99992
3.8	0.99993	0.99993	0.99993	0.99994	0.99994	0.99994	0.99994	0.99995	0.99995	0.99995
3.9	0.99995	0.99995	0.99996	0.99996	0.99996	0.99996	0.99996	0.99996	0.99997	0.99997
4.0	0.99997	0.99997	0.99997	0.99997	0.99997	0.99997	0.99998	0.99998	0.99998	0.99998

TABLE 7.2 Negative Z-values (normalized area to the *left*, where total area = 1)

z	0	0.01	0.02	0.03	0.04	0.05	0.06	0.07	0.08	0.09
-4.0	0.00003	0.00003	0.00003	0.00003	0.00003	0.00003	0.00002	0.00002	0.00002	0.00002
-3.9	0.00005	0.00005	0.00004	0.00004	0.00004	0.00004	0.00004	0.00004	0.00003	0.00003
-3.8	0.00007	0.00007	0.00007	0.00006	0.00006	0.00006	0.00006	0.00005	0.00005	0.00005
-3.7	0.00011	0.00010	0.00010	0.00010	0.00009	0.00009	0.00008	0.00008	0.00008	0.00008
-3.6	0.00016	0.00015	0.00015	0.00014	0.00014	0.00013	0.00013	0.00012	0.00012	0.00011
-3.5	0.00023	0.00022	0.00022	0.00021	0.00020	0.00019	0.00019	0.00018	0.00017	0.00017
-3.4	0.00034	0.00032	0.00031	0.00030	0.00029	0.00028	0.00027	0.00026	0.00025	0.00024
-3.3	0.00048	0.00047	0.00045	0.00043	0.00042	0.00040	0.00039	0.00038	0.00036	0.00035
-3.2	0.00069	0.00066	0.00064	0.00062	0.00060	0.00058	0.00056	0.00054	0.00052	0.00050
-3.1	0.00097	0.00094	0.00090	0.00087	0.00084	0.00082	0.00079	0.00076	0.00074	0.00071
-3.0	0.00135	0.00131	0.00126	0.00122	0.00118	0.00114	0.00111	0.00107	0.00104	0.00100
-2.9	0.00187	0.00181	0.00175	0.00169	0.00164	0.00159	0.00154	0.00149	0.00144	0.00139
-2.8	0.00256	0.00248	0.00240	0.00233	0.00226	0.00219	0.00212	0.00205	0.00199	0.00193
-2.7	0.00347	0.00336	0.00326	0.00317	0.00307	0.00298	0.00289	0.00280	0.00272	0.00264
-2.6	0.00466	0.00453	0.00440	0.00427	0.00415	0.00402	0.00391	0.00379	0.00368	0.00357
-2.5	0.00621	0.00604	0.00587	0.00570	0.00554	0.00539	0.00523	0.00508	0.00494	0.00480
-2.4	0.00820	0.00798	0.00776	0.00755	0.00734	0.00714	0.00695	0.00676	0.00657	0.00639
-2.3	0.01072	0.01044	0.01017	0.00990	0.00964	0.00939	0.00914	0.00889	0.00866	0.00842
-2.2	0.01390	0.01355	0.01321	0.01287	0.01255	0.01222	0.01191	0.01160	0.01130	0.01101
-2.1	0.01786	0.01743	0.01700	0.01659	0.01618	0.01578	0.01539	0.01500	0.01463	0.01426
-2.0	0.02275	0.02222	0.02169	0.02118	0.02068	0.02018	0.01970	0.01923	0.01876	0.01831
-1.9	0.02872	0.02807	0.02743	0.02680	0.02619	0.02559	0.02500	0.02442	0.02385	0.02330
-1.8	0.03593	0.03515	0.03438	0.03362	0.03288	0.03216	0.03144	0.03074	0.03005	0.02938
-1.7	0.04457	0.04363	0.04272	0.04182	0.04093	0.04006	0.03920	0.03836	0.03754	0.03673
-1.6	0.05480	0.05370	0.05262	0.05155	0.05050	0.04947	0.04846	0.04746	0.04648	0.04551
-1.5	0.06681	0.06552	0.06426	0.06301	0.06178	0.06057	0.05938	0.05821	0.05705	0.05592
-1.4	0.08076	0.07927	0.07780	0.07636	0.07493	0.07353	0.07215	0.07078	0.06944	0.06811
-1.3	0.09680	0.09510	0.09342	0.09176	0.09012	0.08851	0.08692	0.08534	0.08379	0.08226
-1.2	0.11507	0.11314	0.11123	0.10935	0.10749	0.10565	0.10383	0.10204	0.10027	0.09853
-1.1	0.13567	0.13350	0.13136	0.12924	0.12714	0.12507	0.12302	0.12100	0.11900	0.11702

-1.0	0.15866	0.15625	0.15386	0.15151	0.14917	0.14686	0.14457	0.14231	0.14007	0.13786
-0.9	0.18406	0.18141	0.17879	0.17619	0.17361	0.17106	0.16853	0.16602	0.16354	0.16109
-0.8	0.21186	0.20897	0.20611	0.20327	0.20045	0.19766	0.19489	0.19215	0.18943	0.18673
-0.7	0.24196	0.23885	0.23576	0.23270	0.22965	0.22663	0.22363	0.22065	0.21770	0.21476
-0.6	0.27425	0.27093	0.26763	0.26435	0.26109	0.25785	0.25463	0.25143	0.24825	0.24510
-0.5	0.30854	0.30503	0.30153	0.29806	0.29460	0.29116	0.28774	0.28434	0.28096	0.27760
-0.4	0.34458	0.34090	0.33724	0.33360	0.32997	0.32636	0.32276	0.31918	0.31561	0.31207
-0.3	0.38209	0.37828	0.37448	0.37070	0.36693	0.36317	0.35942	0.35569	0.35197	0.34827
-0.2	0.42074	0.41683	0.41294	0.40905	0.40517	0.40129	0.39743	0.39358	0.38974	0.38591
-0.1	0.46017	0.45620	0.45224	0.44828	0.44433	0.44038	0.43644	0.43251	0.42858	0.42465
0.0	0.50000	0.49601	0.49202	0.48803	0.48405	0.48006	0.47608	0.47210	0.46812	0.46414

On the normal distribution, μ is shown on the x-axis, and the standard variation is a measure of the data spread. There are two scales, x and z. The x is the scale of your actual parameter (e.g., height), but z is a *normalized* parameter where the mean is centered at zero. To read the standard distribution tables in Table 7.1 and 7.2, x must always be converted to the z scale. Then use z as the normalized value.

The relationship between these variables is as follows:

$$z = \frac{x - \mu}{\sigma}$$

The tables gives $\Phi(z)$, which is the probability that a normally distributed variable will be less than the stated value of z. This is the same as the area under the curve to the *left* of z.

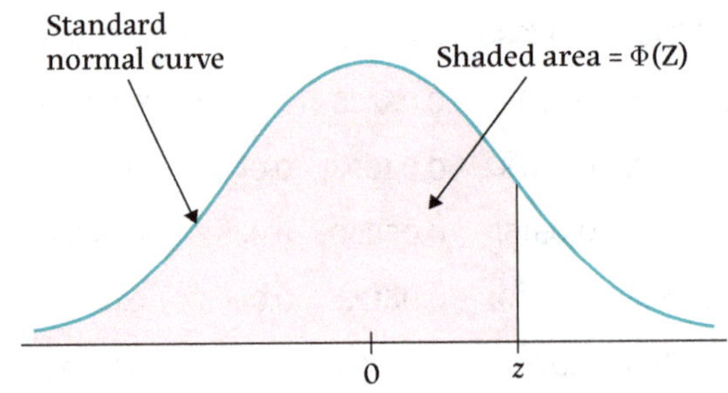

EXAMPLE #2

- What is $\Phi(-1.0)$?

  $\Phi(-1.0) = 0.15866$ (There is a 15.866% chance that this variable will be less than –1.0.)

- What is $\Phi(0)$?

  $\Phi(0) = 0.5000$ (There is a 50% chance that this variable will be less than 0.)

- What is $\Phi(0.25)$?

  $\Phi(0.25) = 0.59871$ (There is a 59.871% chance that this variable will be less than 0.25.)

- What is $\Phi(2.92)$?

  $\Phi(2.92) = 0.99825$ (There is a 99.825% chance that this variable will be less than 2.92.)

EXAMPLE #3

What is the probability that a standard normal variable will be between 1.25 and 2.25?

(Hint: This is the probability that it is less than 2.25 minus the probability that it is less than 1.25.)

$$P(1.25 < z < 2.25) = F(2.25) - F(1.25)$$

$$P(1.25 < z < 2.25) = 0.98778 - 0.89435$$

$$P(1.25 < z < 2.25) = 0.09343$$

$$P(1.25 < z < 2.25) = \mathbf{\underline{9.343\%}}$$

EXAMPLE #3, USING REAL DATA

A 2006 Centers for Disease Control study of 4,482 American males found the mean height = 1763 mm (5 ft 9.4 in), with $\sigma = 114$ mm (4.5 in).

A male at $+2.25\sigma$ is $1763 + 2.25(114) = 2020$ mm (6 ft 7.5 in tall). He is taller than 98.778% (4,427) of these males. Or, we can say that only 55 are taller than him!

A male at $+1.25\sigma = 1763 + 1.25(114) = 1906$ mm (6 ft 3.0 in). He is taller than 89.435% (4,008) of these males.

Also, 9.343% are between these height ranges = 0.09343(4482) = 419 individuals in the study.

Standard Deviations of Normal Curves

As shown in the figure below, approximately 68.3% of data in a normal distribution will fall within one standard deviation of the mean ($\pm 1\sigma$). Approximately 95.5% falls within two standard deviations ($\pm 2\sigma$). About 99.7% falls within three standard deviations ($\pm 3\sigma$).

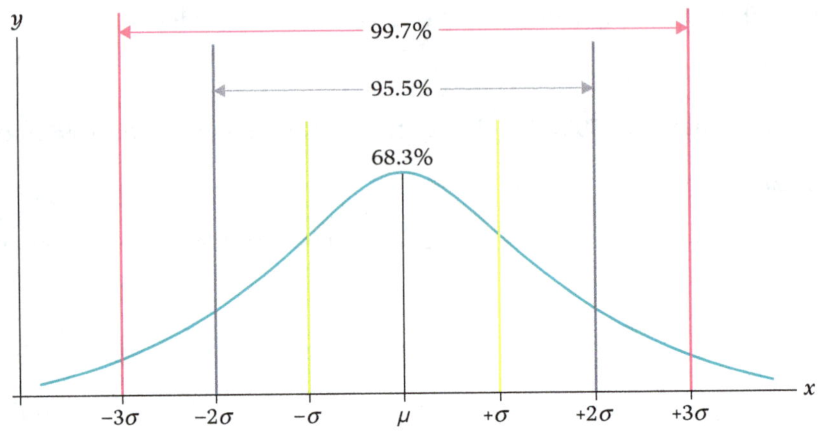

Using the normal distribution, the following can be inferred for the entire population of American males:

- 68.3% of the population is between $\pm 1\sigma$. Thus, 1763 ± 114 mm → Range: 1649 mm (5 ft 4.9 in) to 1877 mm (6 ft 1.9 in).

- 95.5% of the population is between $\pm 2\sigma$. Thus, 1763 $\pm 2(114)$ mm → Range: 1535 mm (5 ft 0.4 in) to 1991 mm (6 ft 6.4 in).

- 99.7% of the population is between $\pm 3\sigma$. Thus, 1763 $\pm 3(114)$ mm → Range: 1421 mm (4 ft 7.9 in) to 2105 mm (6 ft 10.9 in).

Overlap of Two Normal Functions

The difference between two normal functions will also be a normal function. The equations are helpful for comparing the two normal functions, subscripted R for *right* (generally larger values) and L for *left* (generally smaller values), as shown in the figures below. The most interesting thing about this difference is when it is less than zero. This represents the area of overlap, where the "smaller" function has a chance to actually be higher than the "larger" function. For many data sets, R is usually larger than L, but at the overlap, L is larger.

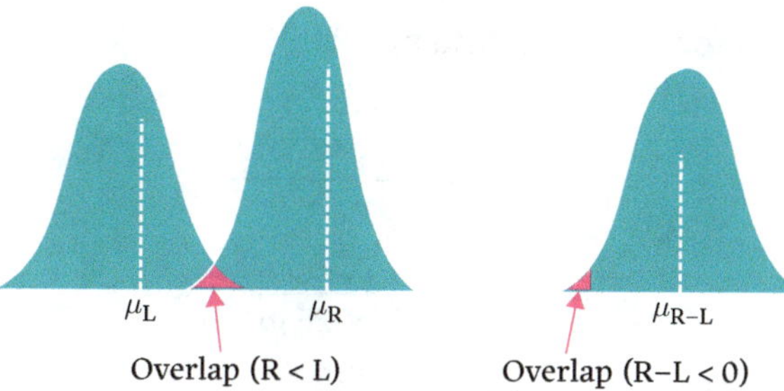

The mean of the difference will be the difference of the means:

$$\mu_{R-L} = \mu_R - \mu_L$$

The standard deviation of the difference can be found by:

$$\sigma_{R-L} = \sqrt{\sigma_R^2 + \sigma_L^2}$$

Why would this be of interest? Imagine if the function R represents the strength (or *resistance*) of a structural beam and the function L represents the *load* applied to that beam. Normally, $R > L$, and everything is fine. However, the probability of overlap is the probability that $R < L$, which would cause the beam to fail. Similar problems arise in the current carried by a conductor or the strength of a bolt.

EXAMPLE #4

The breaking resistance for a concrete column can be modeled as a normal variable R with μ_R = 100,000 lb and σ_R = 10,000 lb. The load applied to the column can be modeled as a normal variable L with μ_L = 75,000 lb and σ_L = 25,000 lb. The difference of these two variables is the *surplus* strength of the column. If it is zero or less, the column may break.

Calculate μ_{R-L}:

$$\mu_{R-L} = \mu_R - \mu_L$$
$$\mu_{R-L} = 100{,}000\,\text{lb} - 75{,}000\,\text{lb}$$
$$\mu_{R-L} = 25.000\,\text{lb}$$

Calculate σ_{R-L}:

$$\sigma_{R-L} = \sqrt{\sigma_R^2 + \sigma_L^2}$$
$$\sigma_{R-L} = \sqrt{(10{,}000\,\text{lb})^2 + (25{,}000\,\text{lb})^2}$$
$$\sigma_{R-L} = 26{,}925\,\text{lb}$$

Convert to standard normal for X = 0 and find probability:

$$Z_{R-L} = \frac{X_{R-L} - \mu_{R-L}}{\sigma_{R-L}}$$

$$Z_{R-L} = \frac{0 - 25{,}000\,\text{lb}}{26{,}925\,\text{lb}}$$

$$Z_{R-L} = -0.93$$

$$\Phi(-0.93) = 0.1762$$

That is a 17.62% chance of failure, an unacceptably high failure rate that could cause injury or even fatality for the people that use engineered products and structures. We usually want this to be literally less than one chance in a million!

Coefficient of Determination

In statistics, the *coefficient of determination* (R^2) is used in the context of statistical models whose main purpose is the prediction of future outcomes, such as with a *linear regression* (i.e., a linear best-fit model). It is the proportion of variability in a data set that is accounted for by the statistical model. It provides a measure of how well future outcomes are likely to be predicted by the model. The closer the value is to 1, the better the model will predict future outcomes. This is the same R^2 encountered when using Microsoft Excel to determine if a linear, polynomial, or other model fits a data set.

Summary

The *normal distribution,* or bell curve, is a useful model for many data sets encountered in engineering. Descriptive values such as the *mean (average)*, *median*, *variance*, and *standard deviation* (*sample* and *population* variants) can be determined for data that follow a normal distribution. Inferred values can be projected from a sample to a larger population in order to evaluate data trends. The probability of a value or range of values can be determined from a data set that follows a normal distribution.

Assignment

1. Show your work to receive credit for this problem. Consider the following data sample:
   X = {11, 15, 5, 12, 20, –3, 3}

   a. What is the mean of this sample?
   b. What is the median of this sample?
   c. What is the sample standard deviation of this sample?
   d. What is the population standard deviation of this sample?

2. Answer the following questions using the standard normal distribution table.

   a. What is $\Phi(-1.65)$?
   b. What is $\Phi(2.47)$?
   c. What is $P(z < -1.65)$?
   d. What is $P(z > 2.47)$?

3. Answer the following questions by first converting to a standard normal value z and then using the table. (Hint: A sketch of the bell curve with the mean and standard deviations noted may be used to assist with the calculations.)

   a. If $\mu = 130$ and $\sigma = 20$, what is $P(x < 180)$?
   b. If $\mu = -25$ and $\sigma = 4$, what is $P(x < -28)$?
   c. If $\mu = 12$ and $\sigma = 1$, what is $P(x > 14.5)$?

4. The breaking resistance (strength) for a concrete column can be modeled as a normal variable R with $\mu_R = 140{,}000$ lb and $\sigma_R = 18{,}000$ lb. The load applied to the column can be modeled as a normal variable L with $\mu_L = 70{,}000$ lb and $\sigma_L = 17{,}000$ lb. What is the probability that a particular load will cause a particular column to fail?

Figure Credits

Probability

Objectives

- Introduce students to probability theory and its application to solving problems such as the classic *birthday problem*.
- Discuss two methods of counting:
  - *Permutations*
  - *Combinations*
- Provide a framework to compute the expectations for a discrete random variable.
- Apply binomial distribution (i.e., determine the number of x heads when tossing y coins).

Introduction to Probability

One of the most commonly applied branches of mathematics is probability. *Probability* determines the likelihood of an event happening. This is assigned a number between 0 and 1. Applications of probability are very evident in everyday life beyond the scope of the sciences. For example, a person cannot turn on the television without the display of chosen lottery numbers for the big prize or the odds that a particular sports team will win the Super Bowl or the World Series. Nevertheless, the discussion of probability in scientific application is important, as it provides an engineer with a methodology to predict the outcome of a given event based on experimental data.

Any discussion of probability begins with the understanding that for any given experiment, an engineer must list all valid and possible outcomes. This list creates what is known as a *sample space,* denoted by the letter S. Distribution of playing cards is a simple application for understanding sample spaces. A card deck has a total of 52 cards. Playing cards are initially divided into two colors—red and black. Each color has two subcategories—hearts and diamonds (red) and spades and clubs (black), totaling 13 cards per subcategory. The 13 cards in each subcategory consist of nine number cards (2–10), three face cards (king, queen, and jack), and one ace. Figure 8.1 shows the sample space for all cards in a deck.

													S
Heart	A	2	3	4	5	6	7	8	9	10	J	Q	K
Space	A	2	3	4	5	6	7	8	9	10	J	Q	K
Club	A	2	3	4	5	6	7	8	9	10	J	Q	K
Diamond	A	2	3	4	5	6	7	8	9	10	J	Q	K

There are two applications of the sample space. First, the sample space can assign an outcome for each item in the sample space, where the total number of outcomes for all items in the sample space is between 0 and 1. In the above sample space, each card is evenly assigned an outcome of 1/52. Second, the sample space can predict a given outcome between items in the sample space. This is known as an *event* (A). An *event* is found within a sample space by computing all total probabilities for items that fit within the event. On the other hand, a *complement* (A′) contains all items that do not fit within an event. The sum of the probability of an event and its complement equals 1.

In summary, the probability of an event

$$P(A) = p(1) + p(2) + p(3).....$$

where

p(1), p(2), p(3) is the probability of an item that fits within an event

Probability

$$P(A) + P(A') = 1$$

Counting Techniques

Further applications of probability can be done with the application of more than one event. The use of sample spaces and the aforementioned equations will come into play in these types of applications.

Another application of probability is the classic *birthday problem*. The birthday problem determines the probability that no two people out of a group of *n* share the same birthday. Consider that there are 365 days within a given year. The first person has a birthday. The second person could have that 1 birthday, or any of the 364 other birthdays. So, the probability that the 2 will not have the same birthday is

$$P(A') = \frac{365}{365} \times \frac{364}{365}$$

If we add a third person, the probability of that person having a birthday different from the other two is

$$P(A') = \frac{365}{365} \times \frac{364}{365} \times \frac{363}{365}$$

and so on.

Recognizing that the numerator multiplies 365 times each successively lower integer as we consider each successive person and the denominator increases by a factor of 365 with each successive person, we can simplify the notation using factorials:

$$n! = n(n-1)(n-2)\ldots(2)(1)$$

Recognize that for the case of 3 persons, the numerator is 365!–362!

So, returning to the question, what is the probability of a group of n people all having a unique birth date?

$$P(A') = \frac{365! - (365-n)!}{365^n}$$

For example, in a group of five people, the probability that no two people share the same birthday is (365 × 364 × 363 × 362 × 361× 360/(365^5))= 0.9729. However, as the number of people in the group increases, the chances that no two people share the same birthday decreases. To apply the birthday problem to more complex situations, we will need to apply other techniques, discussed below.

Probability theory recognizes two major counting techniques—permutations and combinations. A *permutation* is concerned with the order of items chosen; with a *combination,* the order does not matter. Depending on the situation, a permutation or combination can have both repeating and nonrepeating numbers. The numbers that comprise your home address is a permutation because the order of those numbers matters, especially when delivering mail (have you ever received someone else's mail before?). On the other hand, winning games such as bingo or collecting prize money from the lottery is not concerned with the order of the numbers drawn, and these are examples of combinations.

Computing permutations and combinations requires the use of *factorials*. If we assume that n represents any number, then n factorial (notated as $n!$) = $n \times (n-1) \times (n-2) \ldots 1$. For example, $4! = 4 \times 3 \times 2 \times 1 = 24$. Therefore, the calculation of a permutation of n items chosen from a total of r items equals

$$P(r, n) = r!/(r-n)!$$

while a combination of n items chosen from a total of r items equals

$$C(r, n) = r!/[n!(r-n)!]$$

A combination lock uses numbers between 0 and 9. If the lock can use only *four* nonrepeating numbers, how many possible codes can be generated?

Solution:

A combination lock is a permutation. Therefore:

$$P(10,4) = (10!)/[(10-4)!] = 3628800/720 = 5040$$

This states that there are 5,040 codes that can be created if a combination lock uses only four nonrepeating numbers. This number would change if a number can be reused in the permutation.

The lottery allows for a participant to choose three numbers between one and sixty. How many combinations of numbers can a person choose?

Solution:

$$C(60,3) = (60!)/[3!(60-3)!] = 8.32 \times 10^{81}/2.43 \times 10^{77} = 34220$$

This states that there are 34,220 different combinations a person can choose of three numbers between one and sixty. Using counting techniques and probability theory, a person can determine their probability of winning the lottery.

Binomial Distribution

Sometimes it is necessary to predict the outcome of a situation that has only two outcomes. These situations are known as *Bernoulli trials*. Tossing a fair coin would be classified as a Bernoulli trial, since it has only two realistic outcomes—heads or tails. In many cases, one is interested in determining the probability of completing *independent* Bernoulli trials (i.e., each trial is unique), where the number of successful trials is the point of interest. The determination of this probability applies the use of the *binomial distribution*. The binomial distribution uses the following equation:

$$P(X = x) = \binom{n}{x} p^x (1-p)^{n-x}$$

where

    X = a random variable
    x = the number of successful trials
    n = the total number of trials
    p = the probability of a given situation

$$\binom{n}{x} = \frac{n!}{x!(n-x)!}$$

Another important distinction when using binomial distribution is in the interpretation of the wording of a problem. The three main types of binomial distribution problems use one of the following words or phrases: *exactly*, *no more*, and *at least*.

Before attempting to solve a problem using one of these three words, it is important to consider a practical application. For example, if the instructor in a class asks the students, "How many of you have *exactly* five dollars in your pocket?" only the students that have $5.00 in their pockets would raise their hands. This is different than "How many of you have *no more* than five dollars?" This means students with no money in their pockets through students that have exactly $5.00 in their pocket would raise their hands. Finally, if the instructor asks about the number of students that have *at least* five dollars, students that have $5.00 or more in their pocket would raise their hands. In summary, the wording used in the question will determine the probabilities (P(X = x), P(X = x + 1)...) that need to be considered when solving the problem.

EXAMPLE #3

A fair die is rolled five times. What is the probability of generating *no more* than two threes (P(X $\leq$ 2))?

Solution:

Rolling a three on a fair die has a probability of 1/6 (there are six sides to a die). In addition, there are three conditions—the probability of not rolling a three (P(X = 0)), the probability of rolling a three (P(X = 1)), and the probability of rolling two threes (P(X = 2)).

$$P(X=x)=\binom{n}{x}p^x(1-p)^{n-x}$$

$$P(X\leq 2)=P(X=0)+P(X=1)+P(X=2)$$

$$P(X=0)=\binom{5}{0}\left(\frac{1}{6}\right)^0\left(1-\left(\frac{1}{6}\right)\right)^5=0.40188$$

$$P(X=1)=\binom{5}{1}\left(\frac{1}{6}\right)^1\left(1-\left(\frac{1}{6}\right)\right)^4=0.08038$$

$$P(X=2)=\binom{5}{2}\left(\frac{1}{6}\right)^2\left(1-\left(\frac{1}{6}\right)\right)^3=0.1608$$

$$P(X\leq 2)=.40188+0.08038+0.1608=0.6431$$

When rolling a fair die five times, the probability of having no more than two threes is 0.6431.

Optional Coverage: Random Variables

Random variables are described as being either *discrete* or *continuous*. However, this section will only focus on discrete random variables. The outcomes in an experiment of a *discrete random variable* are assigned a numerical value. The outcome is not necessarily known, or the sum of the outcomes does not necessarily equal a definite value. Let's consider *lottery prize* as a discrete random variable. There are nine different prizes in the random variable—$100 million, $1 million, $10,000, $100, $7, and $4. We can express the outcomes in the sample space of this random variable as

$$S = \{100 \text{ million, } 1 \text{ million, } 10{,}000, 100, 7, 4\}$$

A discrete random variable parameter often calculated is the *mean* or the *expectation* (E(X)) of a discrete random variable. The expectation of a discrete random variable adds the product of the assigned probabilities (p_i) for each outcome (x_i) of the random variable. Using the lottery example, suppose that the probability for each prize is 1 in 175 million for winning $100 million, 1 in 5 million for winning $1 million, 1 in 650,000 for winning $10,000, 1 in 20,000 for winning $100, 1 in 360 for $7, and 1 in 55 for winning $4. The expected value for the lottery prize variable is

$$E(X) = \sum p_i x_i$$

$$E(X) = (10000000)\left(\frac{1}{175000000}\right) + \left(\frac{1}{5000000}\right)(1000000) + \left(\frac{1}{650000}\right)(10000)$$

$$+ \left(\frac{1}{20000}\right)(100) + \left(\frac{1}{360}\right)(7) + \left(\frac{1}{55}\right)(4) = 0.884$$

Following this calculation, we can then calculate the expected winnings by taking the difference between the expected value and the cost to purchase a ticket. If a ticket costs $1, then the expected winnings are a loss of $0.12. This states that if one were to play the lottery, it is more likely he or she will lose money.

Summary

The *probability* of an *event* is the likelihood of an event occurring. An event is a subset of a *sample space*, which lists all possible outcomes. Engineers are primarily concerned with the events in the sample space. *Permutations* and *combinations* are counting techniques that can be used to determine the possible number of outcomes. The outcomes of a *discrete random variable* are assigned numbers. The *mean* of discrete random variables is an *expectation*. *Binomial distribution* evaluates the successes of *independent Bernoulli trials* in an experiment.

Assignment

Break into groups. This activity consists of five different activities. Ensure that the groups in the class are evenly distributed based on the number of activities, in order to avoid having to wait. The activities are the following:

1. Dice
2. Birthdays
3. Playing cards
4. Coins
5. Lottery tickets (if time permits)

Each activity has a portion of a data sheet to be filled out (see next page). Follow the instructions carefully for each section.

References

Hayter, Anthony. *Probability and Statistics for Engineers and Scientists. 2nd ed.* Pacific Grove, CA: Duxbury, 2002.

Oakes, William C., and Les L. Leone. *Engineering Your Future. 8th ed.* New York: Oxford University Press, 2015.

Probability Activity Sheet

NAME ____ DATE ___________________

I. Dice

Write down the number of times a die is to be rolled. ________________

Write down a number between one and six. ________________

Roll the die the number of times stated.

Write down the number of times the die rolls the number selected. ________________

Calculate the probability of achieving the result. ________________

II. Birthdays

Write down your birthday (month, day). ________________

Write down the number of students in your group. ________________

Calculate the probability that no one in your group shares a birthday. ________________

How many students in the group share your birthday? ________________

Ask fellow classmates and the instructor their birthdays.

Does anyone in the class have the same birthday? ________________

Calculate the probability that two people in the class have a birthday on the date listed on the first question. ________________

III. Playing Cards

Write down two face, ace, or subcategory cards. ________________

Calculate the probability of choosing the cards in succession (without replacement) in either order. ________________

Calculate the probability of choosing the same card twice (with replacement). ________________

After shuffling the cards, pick two cards. Do not put your first card back.

How many of your two cards did you pick? ________________

Reshuffle the cards. Pick a new card and then put the original card back in the deck. Pick again.

Did you pick the same card twice? ________________

IV. Coins

Write down the number of coins to be tossed simultaneously. _______________

Toss the coins in the air.

Write down the number of heads from the toss. _______________

Calculate the probability of achieving the result. _______________

Repeat the toss.

Write down the number of heads from the toss. _______________

Calculate the probability of achieving the result. _______________

V. Lottery Tickets (Optional)

To play the Powerball, a participant will choose five *unique* numbers between one and 59 *and* one number between one and 35. During the drawing, five white balls between one and 59 are chosen, along with one red ball between one and 35. To win the lottery, a participant must have all values chosen, but the order of white balls does not need to be in the same order of the values drawn.

Based on the given information, compute the probability of becoming the grand prize winner of the Powerball. (Hint: The probability is 1/(the product of two combinations.))

Choose five numbers between one and 59. Choose one number between one and 35. Write in the spaces below. Note that the final space is for the number between one and 35.

_______________ _______________ _______________ _______________

Ask the instructor for the winning numbers. Numbers are chosen every Wednesday.

Did you win anything? _______________

Calculate the probability of winning the Powerball. _______________

Is the risk worth playing the lottery? _______________

Sample Probability Activity Sheet

NAME <u>Sample</u>   DATE <u>02/23/2015</u>

I. Dice

Write down the number of times a die is to be rolled. **<u>4</u>**

Write down a number between one and six. **<u>5</u>**

Roll the die the number of times stated.

Write down the number of times the die rolls the number selected. **<u>2</u>**

Calculate the probability of achieving the result. **<u>0.1157</u>**

$$P(X=x)=\binom{n}{x}p^x(1-p)^{n-x}$$

$$P(X=2)=\binom{4}{2}\left(\frac{1}{6}\right)^2\left(1-\left(\frac{1}{6}\right)\right)^{4-2}$$

$$P(X=2)=6\left(\frac{1}{6}\right)^2\left(1-\left(\frac{1}{6}\right)\right)^2=0.1157$$

II. Birthdays

Write down your birthday (Month, Day). **<u>July 3</u>**

Write down the number of students in your group. **<u>6</u>**

Calculate the probability that no one in your group shares a birthday. **<u>0.9595</u>**

$$(1 \times 364 \times 3\ 63 \times 362 \times 361 \times 360 \times 359)/(365)^6 = 0.9595$$

How many students in the group share your birthday? **<u>0</u>**

Ask fellow classmates and the instructor their birthdays.

Does anyone have the same birthday? **<u>0</u>**

Calculate the probability that two people in the class have a birthday on the date listed on the first question. **<u>0.0055</u>**

$$P(X=x)=\binom{n}{x}p^x(1-p)^{n-x}$$

$$P(X=2)=\binom{41}{2}\left(\frac{1}{365}\right)^2\left(1-\left(\frac{1}{365}\right)\right)^{41-2}$$

$$P(X=2)=820\left(\frac{1}{365}\right)^2\left(1-\left(\frac{1}{365}\right)\right)^{39}=0.0055$$

III. Playing Cards

Write down two face, ace, or subcategory cards. **Ace, red card**

Calculate the probability of choosing the cards in succession (without replacement) in either order. **4/51**

$$P(\text{Ace, Red Card}) = [P(\text{Ace}) \times P(\text{Red Card})] + [\text{Red card}) \times P(\text{Ace})]$$

$$P(\text{Ace})_1 = 4/52 \qquad P(\text{Red})_2 = 26/52$$

$$P(\text{Red})_1 = 26/51 \qquad P(\text{Ace})_2 = 4/51$$

$$P(\text{Ace, Red Card}) = [(4/52) \times (26/51)] + [(26/52) \times (4/51)] = 4/51$$

(Hint: Probability = number of cards fitting requirement/total number of cards x (number of cards fitting requirement left/(total number of cards − 1))

Calculate the probability of choosing the same card twice (with replacement). **1/2704**

$$P(\text{Ace hearts}) = [(P(\text{Ace hearts}) \times P(\text{Ace hearts}] = (1/52) \times (1/52) = 1/2704$$

After shuffling the cards, pick two cards. Do not put your first card back.

How many of your two cards did you pick? **0**

Reshuffle the cards. Pick a new card and then put the original card back in the deck. Pick again.

Did you pick the same card twice?

IV. Coins

Write down the number of coins to be tossed simultaneously. **10**

Toss the coins in the air.

Write down the number of heads from the toss. **4**

Calculate the probability of achieving the result. **0.205**

$$P(X=x)=\binom{n}{x}p^x(1-p)^{n-x}$$

$$P(X=4)=\binom{10}{4}\left(\frac{1}{2}\right)^4\left(1-\left(\frac{1}{2}\right)\right)^6$$

$$P(X=4)=820\left(\frac{1}{2}\right)^4\left(1-\left(\frac{1}{2}\right)\right)^6=0.205$$

Repeat the toss.

Write down the number of heads from the toss. **6**

Calculate the probability of achieving the result. **0.205**

$$P(X=x)=\binom{n}{x}p^{x}(1-p)^{n-x}$$

$$P(X=6)=\binom{10}{6}\left(\frac{1}{2}\right)^{6}\left(1-\left(\frac{1}{2}\right)\right)^{4}$$

$$P(X=4)=820\left(\frac{1}{2}\right)^{6}\left(1-\left(\frac{1}{2}\right)\right)^{4}=0.205$$

V. Lottery Tickets

To play the Powerball, a participant will choose five *unique* numbers between one and 59 *and* one number between one and 35. During the drawing, five white balls between one and 59 are chosen, along with one red ball between one and 35. To win the lottery, a participant must have all values chosen, but the order of white balls does not need to be in the same order of values drawn.

Based on the given information, compute the probability of becoming the grand prize winner of the Powerball. (Hint: The probability is 1/(the product of two combinations)).

Choose five numbers between one and 59. Choose one number between one and 35. Write in the spaces below. Note that the final space is for the number between one and 35.

27 31 42 18 55 22

Ask the instructor for the winning numbers. Numbers are chosen every Wednesday.

Review the lottery key. Did you win anything? **No**

Calculate the probability of winning the Mega Millions. **1/258,890,850 (For this example, the Mega Millions was used instead of Powerball.)**

Consider the Mega Millions:

Choose five numbers between one and 75 and a field of numbers between one and 15.

There are two combinations we must take into account:

$$C(75,5) = 75!/(5! \times 70!) = 17259390$$
$$C(15,1) = 15!/(1! \times 14!) = 15$$
$$1/(17259390 \times 15) = 258890850$$

Is the risk worth playing the lottery? **No**

Circuits and Ohm's Law

Objectives

- Determine the nature of *electric potential, voltage, current,* and *resistance.*
- Apply *Ohm's Law* for DC resistors.
- Derive the power equations for resistors.
- Perform calculations using Ohm's law and power equations.

Electrical Principles

Electric Charge

If an object has either an overabundance of electrons or the absence of electrons, it will have an *electric charge* associated with it. Too many electrons cause an object to have a negative charge, and too few electrons cause an object to have a positive charge. If the electrons are balanced, that is, there are just as many electrons in the atoms as there are protons, the object has no charge at all. There are many ways to move electrons off of one body and onto another.

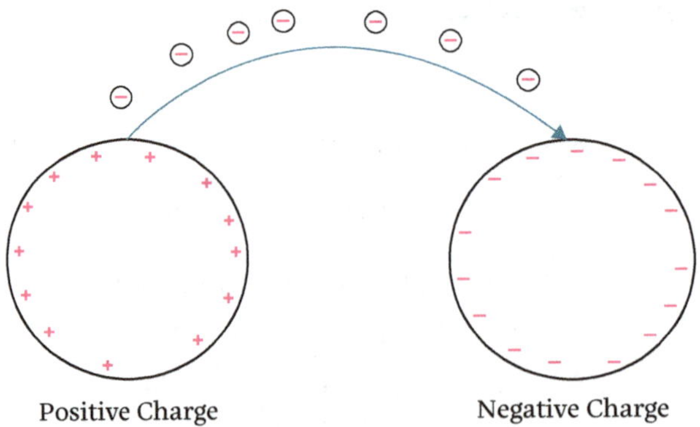

Simply rubbing something can move electrons, such as shuffling your feet across a carpet or sliding across a vinyl seat. If you have ever gotten shocked after sliding across something, you were experiencing a charge. The shock was the rushing of electrons back to where they came from, toward the positive charge.

Electric Potential

You cannot normally see, feel, or taste an electric field (similar to a magnetic field). However, if you have a charged object, it will have an electric field around it. In Figure 9.1, there is an electric field associated with the charged objects. Note that two charged objects with the same charge will repel each other, and two objects with opposite charges will attract each other, again like magnets. It does not take many extra electrons to cause a huge attraction. If two small objects had just 1 percent too many electrons, their repulsion would be about the same as the weight of the earth!

You can see now that if you had a charge (say, something that was negatively charged), it would have a force on it (toward the positive sphere). This force indicates a *potential difference*, or *voltage*. In Figure 9.1, there would be a potential difference (voltage) between the two spheres. The measure of charge has the unit of Coulombs (C). In fact, 1 Coulomb is the charge of 6.242×10^{18} electrons.

Current

If you drop an electron into an electric field, the potential difference (voltage) puts force on it and causes it to move, if it can. The movement of these electrons is called *current*, which is measured in Amperes (sometimes called Amps or just A). One Ampere equals the movement of 1 Coulomb per second.

$$\text{Amperes} = \frac{\text{Coulombs}}{\text{Second}}$$

Although the unit for current is Amps (A), the symbol for current is I.

You would think that the direction of current (electron) flow is in the same direction as the movement of electrons, BUT THIS IS INCORRECT! The direction of current flow is from positive to negative, which is exactly opposite the direction of the flow of electrons. This is because in the early days of electricity study (the 1800s), they thought electricity was a flow of positive particles, so they defined the direction of current from positive to negative. So now, the convention of current flow is from positive to negative.

Resistance

Resistance is just what the word implies, *resistance* to the flow of current. If you have a circuit with a battery and a resistor, the higher the resistor value, the lower the current is going to be. If the resistor is really large, there will only be a trickle of current.

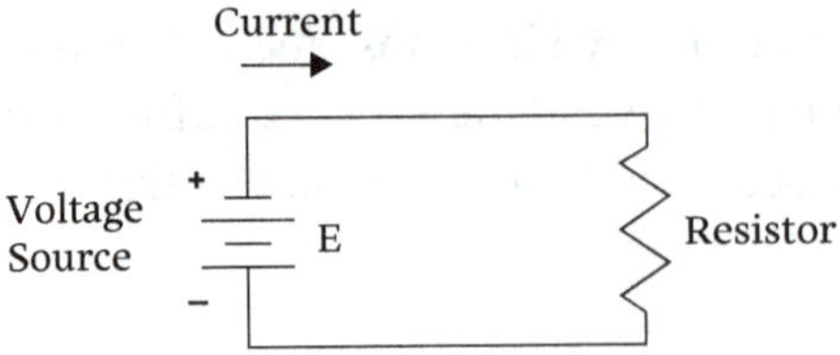

Resistance is measured in Ohms (Ω).

In the real world of electricity, all materials have resistance. Materials that have the least resistance are called *conductors*, and materials that have the greatest resistance are called *insulators*. The type of material determines its resistance. Copper, gold, and aluminum are good conductors and have low resistance; these are used to carry current. Glass, rubber, and some plastics are considered good insulators.

Ohm's Law

As you might guess, the more voltage (force) you have, the more current you will get in a circuit. This can be stated in **Ohm's Law** in the form:

$$\text{Current (A)} = \frac{\text{Voltage (V)}}{\text{Resistance (}\Omega\text{)}} \qquad \text{or simply} \qquad I = \frac{V}{R}$$

$$\text{or solved V} \qquad V = IR \qquad\qquad \text{or solved for R} \qquad R = \frac{V}{I}$$

Where the unit for I (current) is (A), voltage is (V) and resistance is (Ω).

In 1827, a physics professor at the University of Cologne named George Ohm came up with this simple law, which is now called Ohm's law. It was so absurdly simple and hard to prove, theoretically, that he was fired. It took 22 years before it was accepted and he became famous.

This equation says that if you know any two of the three components I, V, or R of a resistor, you can always find the third component. Common units in use for Ohm's law are kilohms (kΩ) and megohms (MΩ), which can be used with milliamps (mA) and microamps (mA). For example, if there is a light bulb with 100 Ω and with a voltage of 120 V, there will be 120/100 = 1.2 A of current through the bulb.

Power

It can be shown that the *power* of a device with resistance is

$$P = I \cdot V, \text{ where P is in Watts, I is in Amperes, and V is in volts}$$

Using Ohm's law to replace I or V, you can also get:

$$P = I^2 \cdot R \text{ or}$$
$$P = V^2/R$$

Notice that if you have any two quantities of the four parameters, P, V, I. and R, you can get the other two from these power equations and Ohm's law. Using the power equations, the power of the components can be determined, such as for appliances. If a device draws 8 A on a 240 volt circuit, the power is 8•240 = 1920 W, or 1.92 kW. Note that energy can be determined by multiplying power by time. So the above device of 1.92 kW, if operated for 10 hours, would use 19.2 kW-hours.

Basic R/C Circuits

Having discussed the relationships between current, voltage, resistance, and power, we can begin to analyze basic circuits.

Kirchoff's Voltage Law

Simplified somewhat, the sum of all the voltage changes around a closed loop must equal zero. Recognize that voltage drops from capacitors may be positive or negative voltage changes, and resistors have a voltage drop in the direction of the current.

$$\sum_{loop} V = 0$$

Consider a simple circuit, like the one in the previous section, which has a capacitor and a resistor. *Capacitor* is a general term for a device that maintains a voltage potential, such as a household battery, although it may be much more sophisticated.

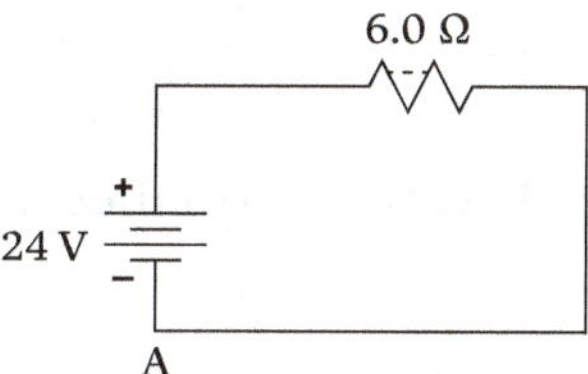

The voltage source is a 24 volt battery, and the resistance is 6.0 W. From the previous section, we can calculate the current and the power to the resistor. We are applying Kirchoff's Voltage law, moving clockwise around the loop, beginning at point A. We can add the 24 V, since the polarity in this direction (indicated by the + and –) is positive for the 24 V capacitor. The voltage drop across the 6.0 W resistor is given by

$$V_{resistor} = -I \times 6.0\ \Omega$$

where the minus sign indicates a voltage drop across the resistor. Since there are no other voltage increases or decreases in the path returning to A, we can use the following equation:

$$+24\ V - I \times 6.0\ \Omega = 0$$

and solve to find

$$I = \frac{24\ V}{6\ \Omega} = 4\ A$$

Can we operate two or more items from a single battery? To answer this question requires a bit more knowledge.

Kirchoff's Current Law

Consider a node within a circuit. This node could be at the junction of multiple parts of the circuit. The sum of all currents into the node must be equal to the sum of all currents out of the node.

$$\sum_{node} I_{in} = \sum_{node} I_{out}$$

Consider two resistors, in parallel, attached to a single voltage supply, as shown in Figure 9.4.

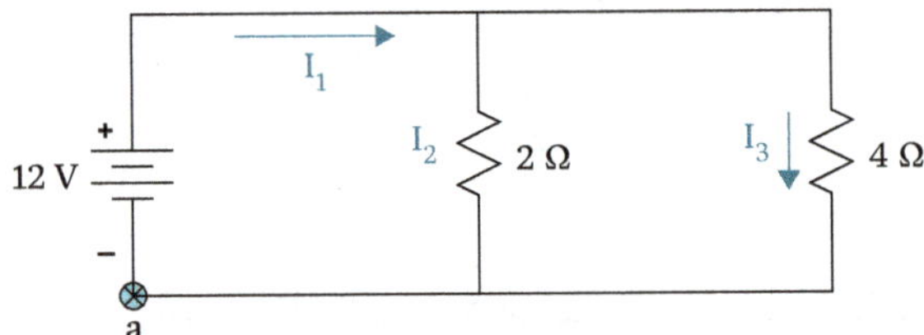

Applying Kirchoff's Current Law to the node directly above the 2 Ω resistor,

$$I_1 = I_2 + I_3$$

Then, applying Kirchoff's Voltage Law to the smaller loop on the left, beginning at point A, located at the lower right,

$$+12V - I_2 \times 2\,\Omega = 0$$

Solving, $I_2 = 6A$
Now, applying the voltage law to the outer loop, again going clockwise from A,

$$+12V - I_3 \times 4\,\Omega = 0$$

Solving, $I_3 = 3A$
And returning to the current law, $I_1 = I_2 + I_3$

$$I_1 = 6\,A + 3\,A = 9\,A$$

Summary

Ohm's Law is a useful method to determine current, voltage, or resistance if you have been given two of the three of a device with resistance. This law also applies to a group of components if their combined resistance can be determined. For example, if a set of combined resistors (some in series, some in parallel) has a known value R, then current to the devices can be determined if the voltage is known. Power equations were also presented so that power can be calculated if two of the three values age given (current, voltage or resistance). Finally, it was shown that one can calculate energy used by multiplying power (kW) times time (hr).

Assignment

1. Find the following unknowns from Ohm's law:

    a. Given a resistor with R = 5 Ω and 60 V, what would the current be?
    b. If a resistor is 20 MΩ and the current is 5 μA, what is the voltage?
    c. If your hair dryer draws 5 A on a 120 V circuit, what is its resistance?

2. What is the power for the hair dryer in Problem 1c above?

3. If you have a resistor rated at 1 W and it has a resistance of 500 Ω, what is the maximum voltage you can apply without overpowering the resistor?

4. There are three appliances plugged into a 120 V outlet. The toaster uses 400 W, the spotlight uses 200 W, and the hair dryer uses 900 W. If the outlet is rated for 15 A, is this an acceptable load for the outlet? Justify your answer by providing calculations.

5. If a central air conditioner uses 20 A at 240 V and operates about 6 hours a day, how many kilowatt-hours does it use in a 30-day month? If the cost of power is $0.10 per kW-hour, how much does it cost to run the AC for one month?

Introduction to Mechanics

Objectives

- Describe mechanics topics typically studied by engineers (*statics*, *dynamics*, and *fluid mechanics*).
- Define the terms *vector* and *scalar*.
- Perform calculations with vectors and scalars.
- Write force vectors in component form in two dimensions.
- Construct a *free body diagram (FBD)* for a rigid body in equilibrium.
- Write and solve equations of *equilibrium*.

Mechanics Defined

Mechanics is the study of the effects of forces on rigid bodies. In the engineering curriculum, there are three foundational mechanics subjects. *Statics* examines rigid bodies at rest or under constant velocity. *Dynamics* concerns rigid bodies subject to unbalanced forces such that they are subject to acceleration due to motion. *Fluid mechanics* examines the effects of rigid bodies subject to air and fluid (i.e., water, oil, and other liquid) pressure. Fluid mechanics incorporates elements of both statics and dynamics, depending on the application. The emphasis in this overview will be on *statics*.

A *rigid body* (i.e., a structure with negligible deformation) is subject to applied force(s) that will, in turn, induce internal forces within the rigid body. A simple *free body diagram* (a diagram with forces acting on the rigid body) of a vise grip is shown in Figure 10.1a. The vise grip is acted upon by the user's hand by a force at the top and bottom of its handles. In turn, the vise grip imparts a force upon an object in the vise grip's jaws. A more intricate free body diagram of a winch is shown in Figure 10.1b.

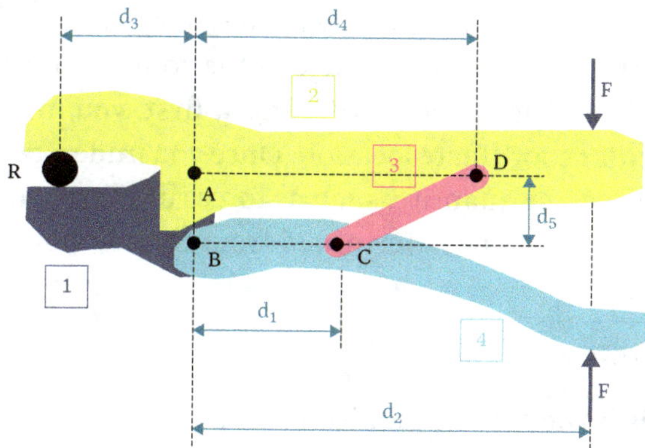

FIGURE 10.1A Vice Grip Free Body Diagram.

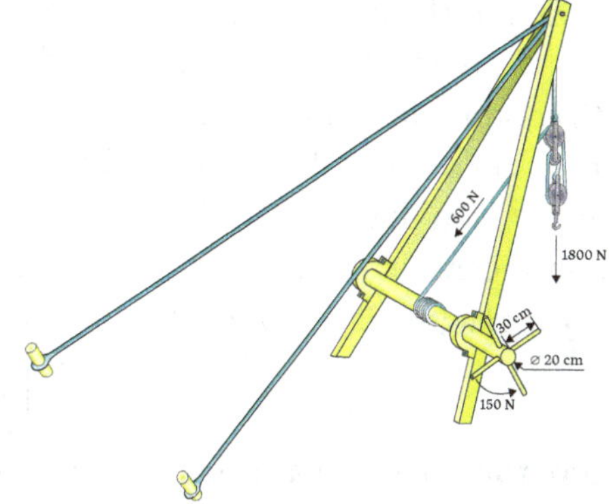

FIGURE 10.1B Winch Free Body Diagram.

Newton's Laws

Newton's three laws are very important in the discussion of mechanics. Newton's first law states that a particle at rest *or* moving in a straight line with constant velocity will remain in this state if it is not subjected to an unbalanced force; this is the case for the study of statics. Newton's second law, *F = ma*, states that force is proportional to mass times acceleration. In the case of statics, the acceleration term due to Earth's gravity, *g*, averages as approximately 32.2 ft/s^2 or 9.81 m/s^2 at the earth's surface. Newton's third law states that the mutual forces of action and reaction between two particles are equal, opposite, and *collinear* (along a line). An example of the third law is that when you stand on the floor, the floor supplies a reactive force against your feet that is equal and opposite to your weight; if the floor cannot supply that force, then you will fall through!

Scalars and Vectors

A *scalar* is a physical quantity having *magnitude* but not associated with a specific direction. Examples of scalars may include mass, length, time, and temperature. A *vector* (written in bold print) is a physical quantity having both *magnitude* and *direction*. Examples of vectors may include displacement, velocity, acceleration, and force.

Vectors are typically described by 1) directional component vectors or 2) a magnitude along with a unit vector. Both methods require us to define a vector of unit length in each coordinate direction. Under the Cartesian system, the unit vectors are defined as:

i or ($\hat{i}$ *or* $\vec{i}$) is a vector in the x-direction with a length of exactly 1 unit.
j or ($\hat{j}$ *or* $\vec{j}$) is a vector in the y-direction with a length of exactly 1 unit.
k or ($\hat{k}$ *or* $\vec{k}$) is a vector in the z-direction with a length of exactly 1 unit.

A vector consisting of directional components has the form

$$\mathbf{A} = A_x\mathbf{i} + A_y\mathbf{j} + A_z\mathbf{k}$$

Where A_x is the magnitude in the x (or $\mathbf{i}$) direction, A_y is the magnitude in the y (or $\mathbf{j}$) direction, and A_z is the magnitude in the z (or $\mathbf{k}$) direction. This requires knowledge of the components in each direction, but not immediately the magnitude. If this sounds confusing at first, you might try to think of these as distances leading to a particular coordinate location. Once you understand the concept, you may extend that knowledge to any vector quantity, such as forces or moments.

A vector consisting of a magnitude and a unit vector may be determined from the component vector. The magnitude of the vector can be found using

$$|A| = \sqrt{A_x^2 + A_y^2 + A_z^2}$$

where the absolute value bars indicate that this is the scalar magnitude of the vector **A**.

A vector of unit length, **u**, is also needed in the particular direction of the initial vector **A**.

$$u = \frac{A_x + A_y + A_z}{|A|}$$

The vector is now expressed as

$$A = |A|u$$

While the uses of this expression may not be immediately clear, it will prove extremely useful in your future mechanics courses.

Example:

Scalar: 50 N (magnitude but no direction)

Vector: 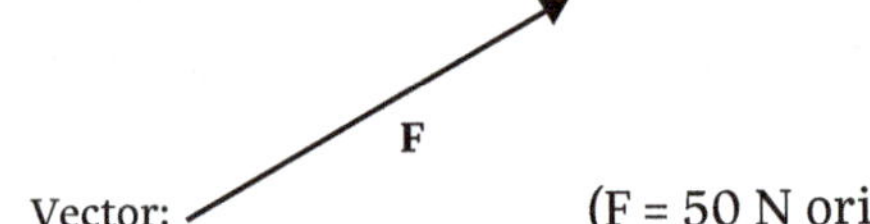

(F = 50 N oriented 30° clockwise from the horizontal x-axis)

Vector Representation

Vectors can be represented graphically, as shown above, or analytically, using *vector components* based on the **i**, **j**, and **k** unit vectors (or x, y, z directions). Using the analytical method, consider Cartesian vector **A**. Vector **A** has a magnitude of A. Vector **A** can be written as $\mathbf{A} = A_x\mathbf{i} + A_y\mathbf{j} + A_z\mathbf{k}$ (3-dimensional case). The magnitude of A = $(A_x^2 + A_y^2 + A_z^2)^{0.5}$ When dealing with two dimensions (2-D), the **k** unit vector and the A_z magnitude component are both *zero*.

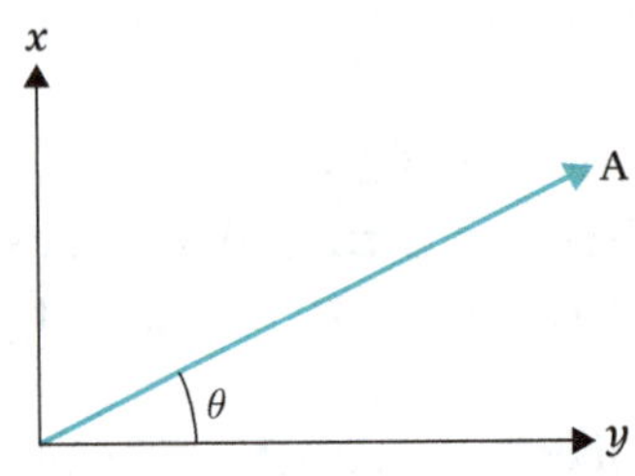

EXAMPLE #1: DETERMINATION OF POSITION, UNIT, AND FORCE VECTORS

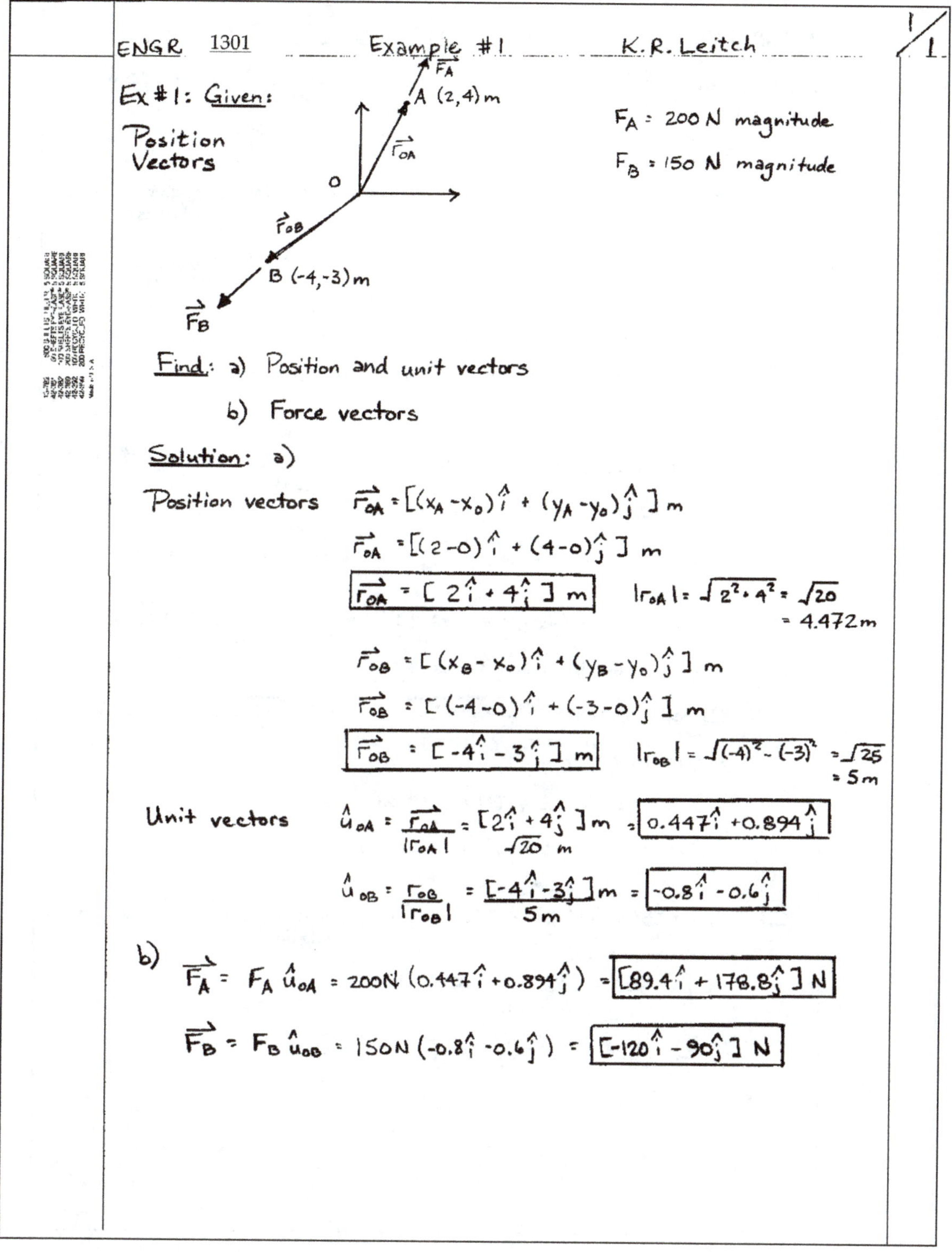

EXAMPLE #2: RESOLVING VECTORS INTO COMPONENTS

ENGR 1301 | Example #2 | K.R. Leitch | 1/1

Ex #2: Given:

$C = 60\overline{0}$ lb

$A = 22\overline{0}$ lb

3 5 4

$74.6°$

$105.4°$

$B = 125$ lb

Hint: SOH CAH TOA

Sine = $\dfrac{\text{opposite}}{\text{hypotenuse}}$

Cosine = $\dfrac{\text{adjacent}}{\text{hypotenuse}}$

Tangent = $\dfrac{\text{opposite}}{\text{adjacent}}$

Find: Resolve each vector into x and y components

Solution:

A: $A_x = 220\text{ lb} \cos 30° = 190.5$ lb $\rightarrow$ Vector Notation $\vec{A} = (190.5\hat{i} + 110\hat{j})$ lb

$A_y = 220\text{ lb} \sin 30° = 110$ lb $\uparrow$

Check $A = 220 = \sqrt{A_x^2 + A_y^2} = \sqrt{190.5^2 + 110^2} = 220$ lb ✓

B: $B_x = 125\text{ lb} \cos(-105.4°) = -33.2$ lb $\leftarrow$ Vector Notation $\vec{B} = (-33.2\hat{i} - 120.5\hat{j})$ lb

$B_y = 125\text{ lb} \sin(-105.4°) = -120.5$ lb $\downarrow$

Check $B = 125 = \sqrt{B_x^2 + B_y^2} = \sqrt{(-33.2)^2 + (-120.5)^2} = 125$ lb ✓

Alternate Method

$B_x = -125\text{ lb}(\cos 74.6°) = -33.2$ lb $\leftarrow$

$B_y = -125\text{ lb}(\sin 74.6°) = -120.5$ lb $\downarrow$

C: $C_x = -600\text{ lb}\left(\dfrac{4}{5}\right) = -480$ lb $\leftarrow$ Vector Notation $\vec{C} = (-480\hat{i} + 360\hat{j})$ lb

$\qquad$ (adj / hyp.)

$C_y = 600\text{ lb}\left(\dfrac{3}{5}\right) = 360$ lb $\uparrow$

$\qquad$ (opp. / hyp.)

Check $C = \sqrt{C_x^2 + C_y^2} = \sqrt{(-480)^2 + (360)^2} = 600$ lb ✓

EXAMPLE #3: VECTOR ADDITION

ENGR 1301   Example #3   K.R. Leitch   1/1

Ex. #3: Vector Addition

Given:

$C = 600 \, lb$

$A = 220 \, lb$

$30°$

$105.4°$

$B = 125 \, lb$

Find: a) the resultant of A and B

b) the resultant of A, B, and C

Solution:

a) $R_{x_1} = A_x + B_x = 190.5 \, lb + (-33.2 \, lb) = 157.3 \, lb \rightarrow$

$R_{y_1} = A_y + B_y = 110 \, lb + (-120.5 \, lb) = -10.5 \, lb \downarrow$

$R_1 = \sqrt{R_{x_1}^2 + R_{y_1}^2} = \sqrt{(157.3)^2 + (-10.5)^2} = \boxed{157.7 \, lb}$

$\Theta_1 = \tan^{-1}\left(\frac{R_{y_1}}{R_{x_1}}\right) = \tan^{-1}\left(\frac{-10.5}{157.3}\right) = \boxed{-3.81°}$

Vector Notation $\quad \vec{R_1} = (157.3\,\hat{i} - 10.5\,\hat{j}) \, lb$

$A = 220 \, lb$

$\Theta = 3.81°$

$B = 125 \, lb$

$R_1 = 157.7 \, lb$

b) $R_{x_2} = A_x + B_x + C_x = 190.5 \, lb + (-33.2 \, lb) + (-480 \, lb) = -322.7 \, lb \leftarrow$

$R_{y_2} = A_y + B_y + C_y = 110 \, lb + (-120.5 \, lb) + 360 \, lb = 349.5 \, lb \uparrow$

$R_2 = \sqrt{R_{x_2}^2 + R_{y_2}^2} = \sqrt{(-322.7)^2 + (349.5 \, lb)} = \boxed{475.7 \, lb}$

$\Theta_2 = \tan^{-1}\left(\frac{R_{y_2}}{R_{x_2}}\right) = \tan^{-1}\left(\frac{349.5}{-322.7}\right) = \boxed{-47.3°}$

Vector Notation $\vec{R_2} = (-322.7\,\hat{i} + 349.5\,\hat{j}) \, lb$

Force Systems and Free Body Diagrams

Types of Forces

Forces that act along the same line of action are *collinear forces*. These forces can be added and subtracted algebraically. Forces that pass through the same point are *concurrent forces*. Forces acting in the same plane are *coplanar forces*. Forces can simultaneously be *concurrent* and *coplanar*.

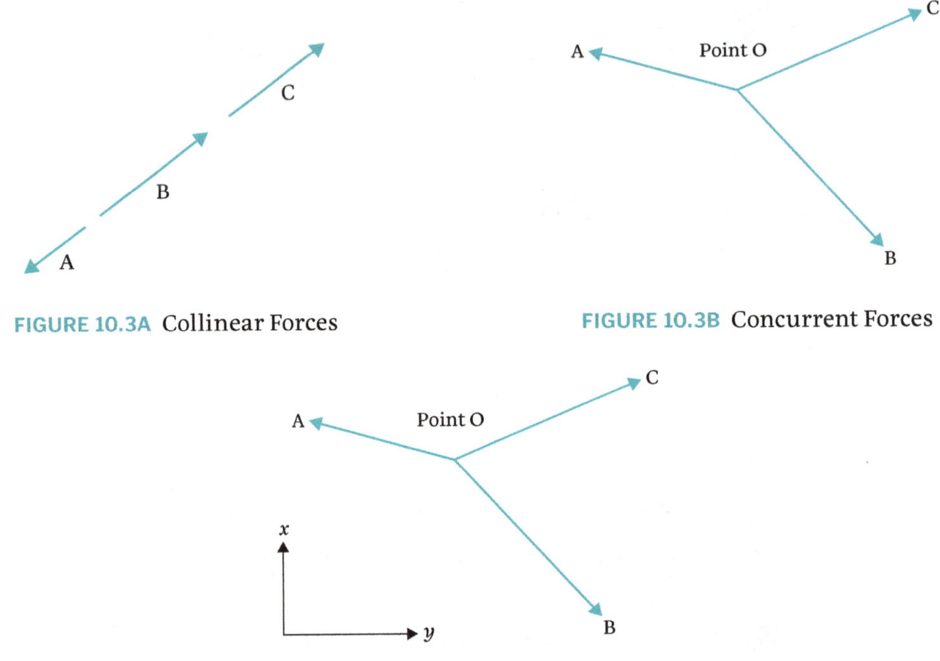

FIGURE 10.3A Collinear Forces

FIGURE 10.3B Concurrent Forces

FIGURE 10.3C Concurrent and Coplanar (x-y plane)

A frictionless pulley redirects the force in the cable as long as no other forces are present. This is the principle of **transmissibility**.

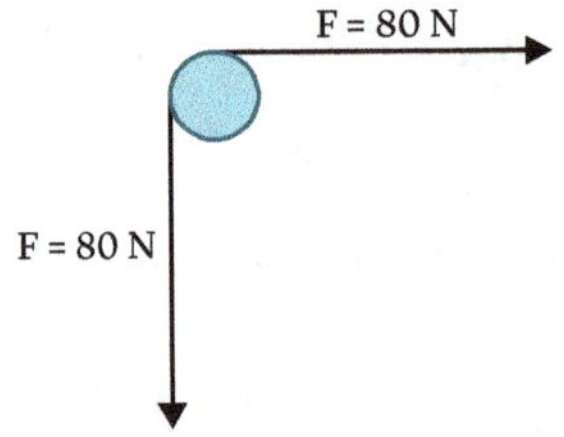

FIGURE 10.4 Frictionless Pulley

Free Body Diagrams (FBDs)

A *free body diagram* (*FBD*) is a sketch of a rigid body with all forces acting on it. The rigid body is isolated from its surroundings. Since Newton's laws apply, we must replace the supports and/or connecting parts with equivalent forces.

There are many kinds of *supports*. Three common types will be mentioned here due to their common uses. A **cable** (or chain or rope) provides a tension (pulling) force only. A **roller** (or wheel or smooth edge) provides a single force perpendicular to a surface. A **hinge** or **pin** provides a force at some angle, normally resolved into x and y components.

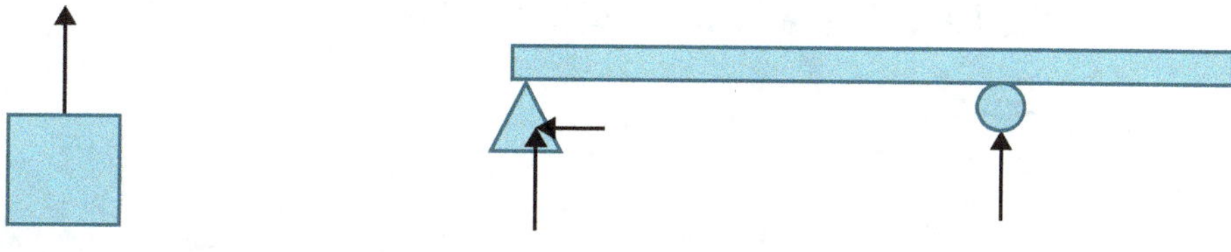

FIGURE 10.5A Cable Support   **FIGURE 10.5B** Hinge (Left) and Roller (Right)

A complete free body diagram has labeled points with all external and support forces noted. Dimensions are commonly given on FBDs for computational purposes.

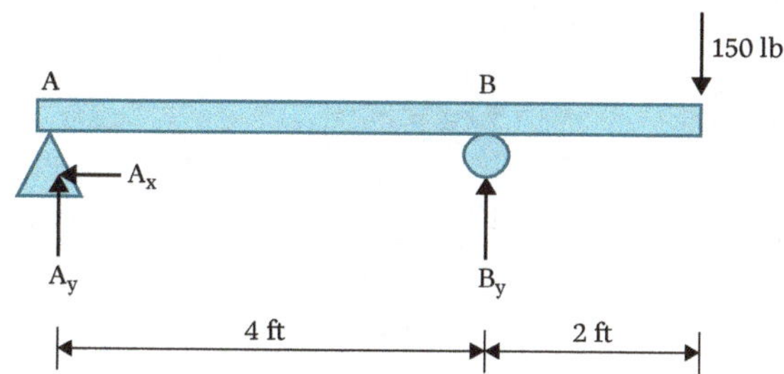

FIGURE 10.6 Example Free Body Diagram of a Diving Board

Equilibrium

Recall Newton's laws. Based on these laws, we can state equilibrium in 2-D as $\sum F_x = 0$, $\sum F_y = 0$, and $\sum M = 0$. For a particle in equilibrium, only the first two equations apply. When finding the resultant of a series of forces on a particle, we can rewrite the top two equations as $\sum F_x = R_x$ and $\sum F_y = R_y$. An example of equilibrium is given on the following two pages.

Summary

Mechanics is the study of the effects of forces on rigid bodies. *Statics* examines rigid bodies at rest or under constant velocity. *Dynamics* concerns rigid bodies subject to unbalanced forces such that they are subject to acceleration due to motion. *Fluid mechanics* examines the effects of rigid bodies subject to air and fluid (i.e., water, oil, and other liquid) pressure. A *rigid body* is a structure with negligible deformation subject to internal and external forces. These forces can be documented using a sketch called a *free body diagram* (FBD). Statics deals with rigid bodies in *equilibrium* with external support forces (e.g., from cables, rollers, or pins) and external forces. Forces can be solved for using *equations of equilibrium*.

A *scalar* is a physical quantity having *magnitude* but no *direction*. A *vector* is a physical quantity having both *magnitude and direction*. Vectors can be written in component form. Vectors can be *collinear* (along one line) or *concurrent* (going through one point). Vectors may also be *coplanar*.

EXAMPLE #4A: PARTICLE EQUILIBRIUM USING A FORCE TABLE

ENGR 1301   Example #4   K.R. Leitch   1/2

Ex #4: Force Table Equilibrium

Given:

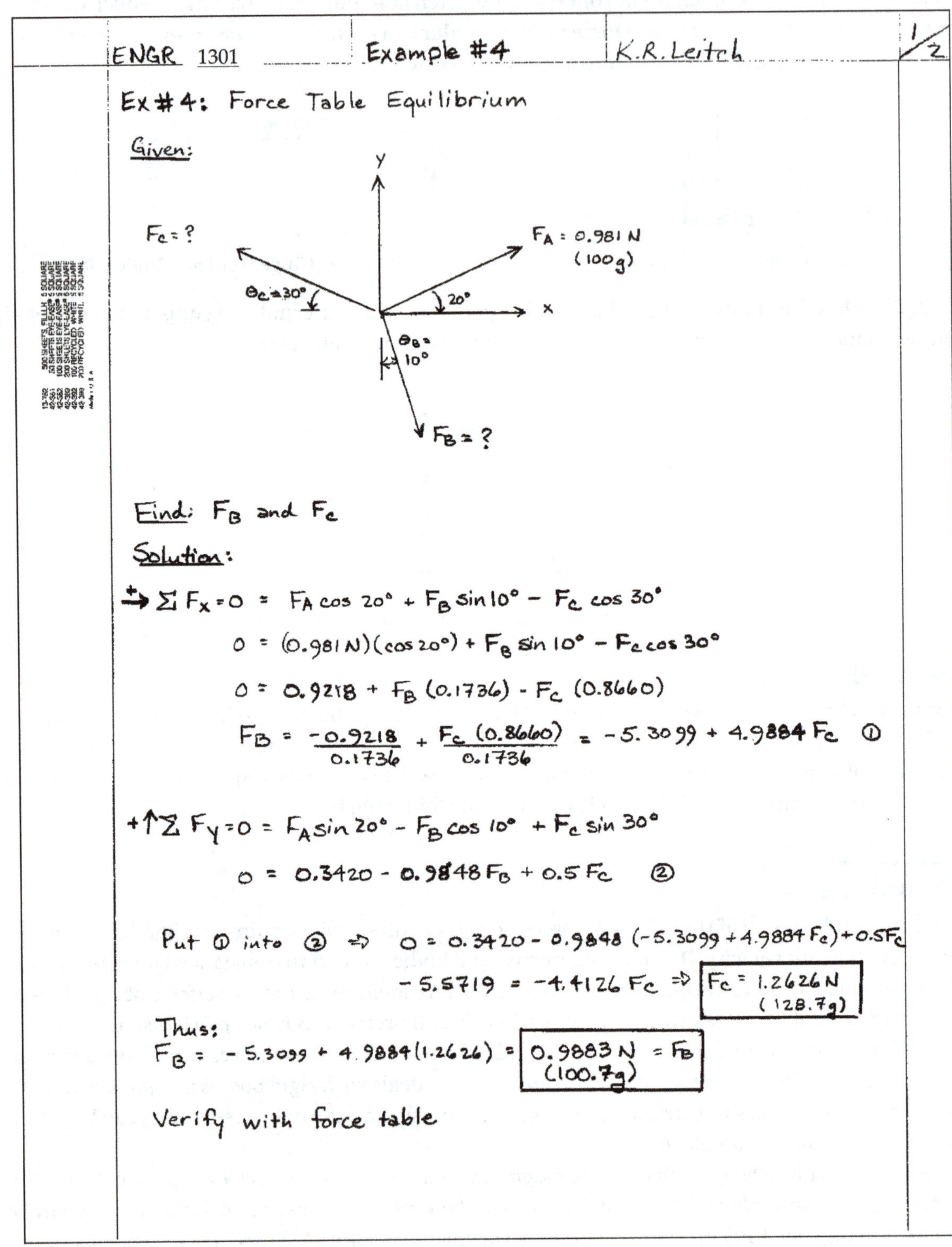

Find: F_B and F_C

Solution:

$$\xrightarrow{+}\ \Sigma F_x = 0 = F_A \cos 20° + F_B \sin 10° - F_C \cos 30°$$

$$0 = (0.981\ N)(\cos 20°) + F_B \sin 10° - F_C \cos 30°$$

$$0 = 0.9218 + F_B (0.1736) - F_C (0.8660)$$

$$F_B = \frac{-0.9218}{0.1736} + \frac{F_C (0.8660)}{0.1736} = -5.3099 + 4.9884\, F_C \quad ①$$

$$+\uparrow \Sigma F_y = 0 = F_A \sin 20° - F_B \cos 10° + F_C \sin 30°$$

$$0 = 0.3420 - 0.9848\, F_B + 0.5\, F_C \quad ②$$

Put ① into ② $\Rightarrow$ $\quad 0 = 0.3420 - 0.9848\,(-5.3099 + 4.9884\, F_C) + 0.5 F_C$

$$-5.5719 = -4.4126\, F_C \Rightarrow \boxed{F_C = 1.2626\ N \\ (128.7\ g)}$$

Thus:
$$F_B = -5.3099 + 4.9884\,(1.2626) = \boxed{0.9883\ N = F_B \\ (100.7\ g)}$$

Verify with force table

EXAMPLE #4B: PARTICLE EQUILIBRIUM USING A FORCE TABLE (ALTERNATE SOLUTION)

ENGR 1301 | Example #4 | K. R. Leitch | 2/2

Alternate Solution Using Matrices

From previous:

$\xrightarrow{+}\ \Sigma F_x = 0 = 0.9218 + F_B\,(0.1736) - F_c\,(0.8660)$

$\qquad$ OR $\quad 0.1736\,F_B - 0.8660\,F_c = -0.9218$

$\xrightarrow{+\uparrow}\ \Sigma F_y = 0 = 0.3420 - 0.9848\,F_B + 0.5\,F_c$

$\qquad$ OR $\quad -0.9848\,F_B + 0.5\,F_c = -0.3420$

Thus:

$$\begin{bmatrix} 0.1736 & -0.8660 \\ -0.9848 & 0.5 \end{bmatrix} \begin{Bmatrix} F_B \\ F_c \end{Bmatrix} = \begin{Bmatrix} -0.9218 \\ -0.3420 \end{Bmatrix}$$

OR $\quad \begin{Bmatrix} F_B \\ F_c \end{Bmatrix} = \begin{bmatrix} 0.1736 & -0.8660 \\ -0.9848 & 0.5 \end{bmatrix}^{-1} \begin{Bmatrix} -0.9218 \\ -0.3420 \end{Bmatrix}$

Yielding $\begin{Bmatrix} F_B \\ F_c \end{Bmatrix} = \begin{Bmatrix} 0.9883 \\ 1.2626 \end{Bmatrix} N$

OR $\quad \boxed{\begin{aligned} F_B &= 0.9883\,N \\ F_c &= 1.2626\,N \end{aligned}}$ $\quad$ Check

Assignment

Given the following with origin O at (0,0):

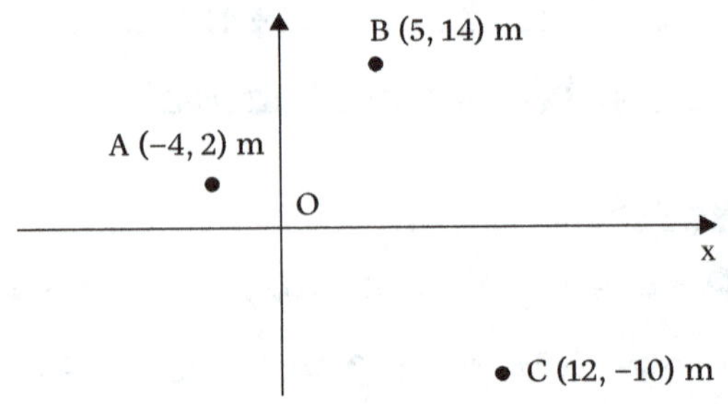

1. Determine the position vectors from O to A ($\mathbf{r}_{OA}$), O to B ($\mathbf{r}_{OB}$), and O to C ($\mathbf{r}_{OC}$).

2. Determine the magnitude of the three position vectors in Problem #1.

3. Determine unit vectors along OA ($\mathbf{u}_{OA}$), OB ($\mathbf{u}_{OB}$), and OC ($\mathbf{u}_{OC}$).

4. Determine a force vector $\mathbf{F}_A$ along the same line of action as the unit vector from O to A ($\mathbf{u}_{OA}$) if the magnitude of the force vector is 300 N.

5. Determine a force vector $\mathbf{F}_B$ along the same line of action as the unit vector from O to B ($\mathbf{u}_{OB}$) if the magnitude of the force vector is 500 N.

6. Determine a force vector $\mathbf{F}_C$ along the same line of action as the unit vector from O to C ($\mathbf{u}_{OC}$) if the magnitude of the force vector is 250 N.

7. Determine a resultant force vector $\mathbf{F}_R$ that is the vector sum of force vectors $\mathbf{F}_A$ and $\mathbf{F}_B$. Find the magnitude and orientation of the resultant force vector $\mathbf{F}_R$.

Figure Credits

Course Labs and Exercises

Scavenger Hunt

Engineering Design: Paper Airplanes

West Point Bridge Designer Lab

Tensile Testing of Materials Lab

Spot Speed Study Lab

Elastic Spring Lab

Microsoft Excel Lab

Launcher Project

Ohm's Law and Circuits Lab

Water Hardness Lab

Scavenger Hunt

OBJECTIVES:

1. Locate several important campus resources for students.

2. Learn to work in a group setting to complete an assignment.

3. Compile a group report according to class requirements in a timely manner.

PROCEDURE:

1. Form a group of two to three students.

2. Visit and photograph the following locations:

   a. Career Services

   b. Library

   c. Computer Lab

   d. Registrar

   e. Business Office

   f. Cafeteria

   g. Activities Center

   h. Engineering Office

   i. Writing Center

   j. Bookstore

   k. Extra credit: up to two additional campus resources

3. One report will be delivered by each group. Students in the same group will receive the same grade. Make sure everyone contributes equally.

DELIVERABLES:

Students will submit one report per group. This report will consist of the following sections:

1. Outside cover (be creative)

2. Title page with group member names, class section, and instructor's name

3. One page for each location, including:

   a. Picture of your group at that location (clearly marked)

   b. What services or resources are provided to students at that location

   c. How your group plans to use these resources during the next year

Make sure to use spell check. Be creative!

Engineering Design: Paper Airplanes

OBJECTIVES:

1. Explore the engineering design process by creating a paper airplane.

2. Document the steps in a concise, informative manner so that another person or group can recreate the paper airplane.

3. Develop a team to perform work and complete a project on time according to specifications.

PROCEDURE:

1. Form a group of two to three students.

2. Part #1 (This week)

   a. Develop a paper airplane within the group from an existing or new design. *Note:* If no one in the group knows a paper airplane design, consult with your instructor for assistance.

   b. Document in words the process to create a copy of the group's airplane (see Part #1 deliverables below).

3. Part #2 (Next week)

   a. Follow another group's instructions.

   b. Provide written feedback for the other group in the form of a business-style memo.

DELIVERABLES:

1. Part #1

   a. Each group will write a summary memo addressing the following:

      i. Briefly describe the design process, including how the group functioned as an entity.

      ii. Write a bulleted list of instructions (approximately ten steps) such that another person or group can follow it to create a copy of the group's airplane design.

   b. Include a working prototype paper model of your airplane. Make sure to label it with the group members' names.

2. Part #2

   a. Provide written feedback to another group in the form of a business memo.

   b. Give the other group constructive feedback and comment on your own group's process of creating the other airplane design from the set of instructions.

Acknowledgment for this activity: Dr. Kenneth Leitch

Source: https://www.okcareertech.org/educators/cimc/resources/resources-by-title/supplemental-resources-for-tlas/filefolderbridge.pdf.

West Point Bridge Designer Lab

OBJECTIVES:

1. Design a single-span truss bridge within a set of specifications, as described in the procedure.

2. Observe the general characteristics of materials in tension and compression.

PROCEDURE:

1. Form a group of two to three students.

2. One design will be delivered by each group. Students in the same group will receive the same grade.

3. Use and/or obtain the latest version of *West Point Bridge Designer* (WPBD) for your operating system. This software can be obtained free of charge at http://bridgecontest.org.

4. The bridge shall be designed to accommodate the following specifications:

    a. The structure will consist of a *single-span truss*, with a level roadway. **BE SURE TO DESIGN THE SINGLE SPAN TRUSS**. You may choose any template or start with a blank screen.

        i. No intermediate piers are permitted.

        ii. Standard or arch abutments may be utilized.

        iii. No cable anchorages may be used.

        iv. Choose *high-strength concrete* and *two AASHTO trucks*.

    b. To provide clearance for trucks, a minimum of 4.0 meters shall be provided between the road surface and the top chord of a through-truss. (Please note that the road surface is not at zero, but rather above zero. See the scale on the side of the screen for elevations.)

    c. To provide clearance for overhead power lines, the height of a through-truss bridge (above the roadway) cannot exceed 7.0 meters.

    d. The program will permit you to put nodes and members at the flood stage of the river. However, you are required to maintain a minimum of 6.0 meters of clearance between the flood stage and the bottom of your truss at midspan (equal distance between the two abutments).

DELIVERABLES:

1. **Design**: An engineer's primary responsibility is to deliver a safe design that can be constructed within budget *in addition to* satisfying functional, strength, and stiffness criteria. Your design will be graded on being within budget and having sufficient strength. However, all designs within budget are *not* created equal. Cost figures for material, labor, and products are included in the WPBD.

2. **Memo**: Write a brief memo describing your bridge design, how it was created, major problems/discoveries, major conclusions, the price of your bridge design, and the members and materials selected and why.

3. **Extra Credit**: Design a bridge as similar as possible to your first bridge, but with standard concrete. In your memo, compare the bridge costs using high-strength concrete and ordinary concrete. How did member size change due to the increased weight of the ordinary concrete?

An inadequately designed deck truss example (too expensive):

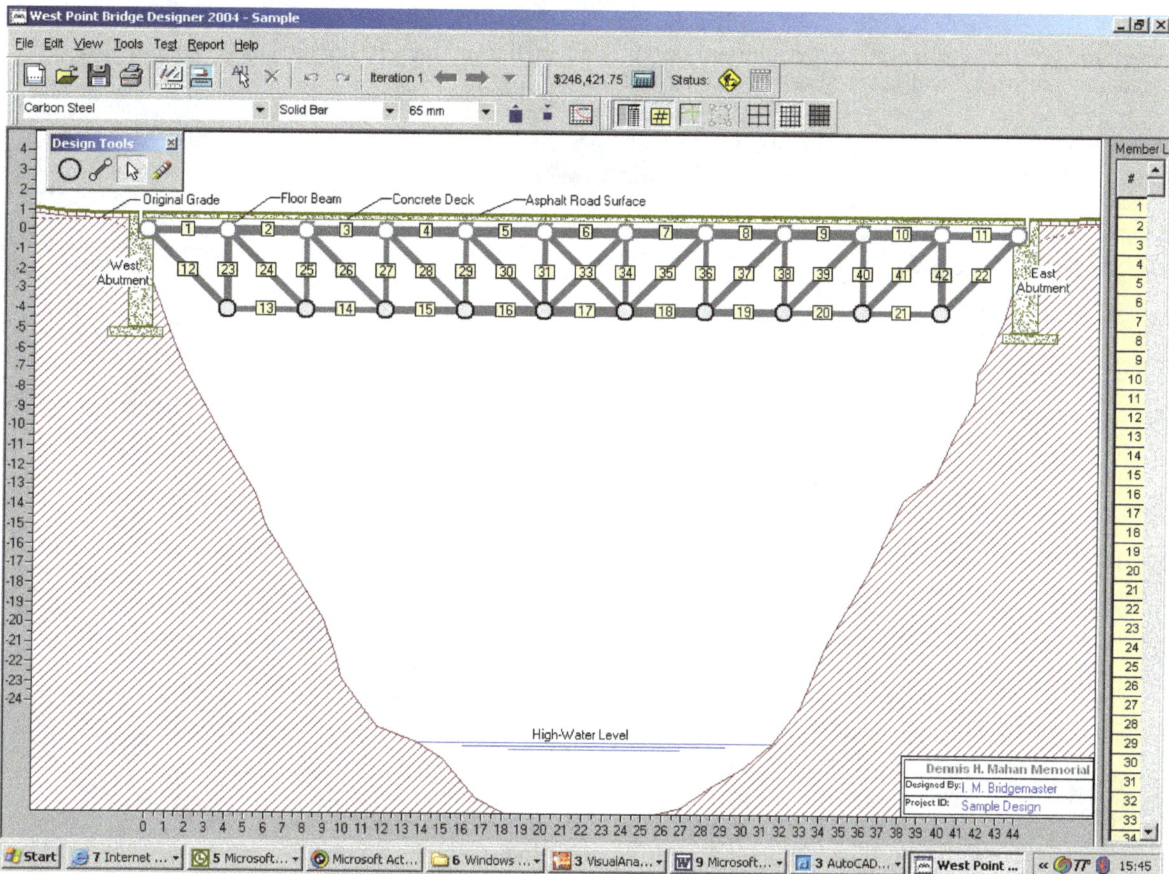

WAYS TO IMPROVE YOUR DESIGN:

1. Use solid bars for tension members.

2. Use tubular instead of solid members for compression members.

3. Use as many of the same type of members as possible to reduce the product cost.

4. See the help section of the WPBD.

Source: https://www.okcareertech.org/educators/cimc/resources/resources-by-title/supplemental-resources-for-tlas/filefolderbridge.pdf.

5. When a very long member fails in compression, try replacing it with two members with a joint in the middle; you'll also need to put in a third member that braces this member. (This reduces the effective length of the compression member and thereby can significantly increase the compressive strength.)

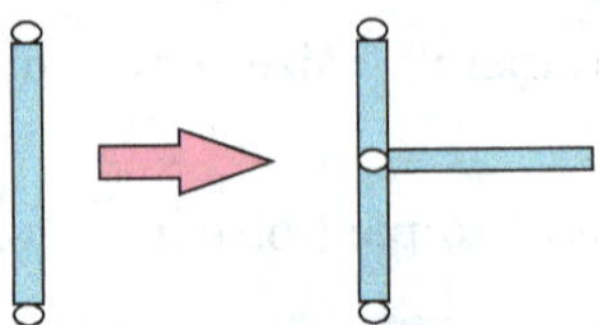

Acknowledgment for this activity: Dr. Kenneth Leitch, as adapted from materials created by West Point Military Academy. See https://bridgecontest.org for more information.

Tensile Testing Lab

OBJECTIVES:

- Test a material in tension using the concept of mechanical advantage with a lever arm (depicted in the figure below, built of common materials).
- Determine the relationship between the width of the material specimens and tensile strength.

FIGURE A.1 Lever Arm Apparatus

PROCEDURE:

1. Form a group of two to three students.

2. One report will be delivered by each group. Students in the same group will receive the same grade. Make sure everyone contributes equally.

3. Each group will test one width of tension specimen. The specimens are cut perpendicular to the long edge of a common manila paper file folder (a paper cutter is recommended for uniform cuts to this thick paper material). Each group will test _______ specimens. All groups will share the data to fill the table provided below.

4. Students can use predetermined weights or sand (of unknown but measurable mass) to test the tension specimens using a lever arm as denoted in the figure in this lab. (This lever arm was built with a mechanical advantage of 4:1, the ratio of L2/L1 in this lab.) A weighing scale can be used to determine the mass (weight) used to break the specimens.

5. Students will create a table for the test measurements they conduct on each width of tensile specimen.

6. All units are to be in SI (metric) units for all results.

7. Students should follow the guidelines for writing technical reports found at the end of the text.

RESULTS:

For this lab, the following results are required:

1. Table showing the tensile test measurements.

2. Figure that shows the tensile strength (N) on the y-axis, as a function of tensile specimen width on the x-axis. Sketch a best-fit line through the data points, including the point (0,0). (This means that at zero width, there is zero strength. On the trendline, force the intercept through zero.)

3. Estimation of the tensile strength (N) for the following widths:

    a. 5 mm

    b. 7 mm

    c. 9 mm

    d. 11 mm

    e. 13 mm

EQUATIONS TO BE USED:

Tensile Strength:

$$T = \frac{WL_2}{L_1} \tag{1}$$

where

    T = Tensile strength (N)
    L_1 = Length from center of tension specimen to pivot
    L_2 = Length from pivot to center of notch where weight is located (mm)
    W = Weight at notch

$$W = mg \tag{2}$$

where

    m = mass (kg)
    g = gravitational acceleration (= 9.81 m/s^2)

Note: 1 kg*m/s^2 = 1 N (Newton)

DELIVERABLES:

Students shall submit one report per group. This report shall consist of the following sections:

1. **Cover Page**: The report title page should contain the name of the laboratory experiment, date submitted, class section, and all student group members' names.

2. **Abstract**: Clearly state in approximately two paragraphs a summary of the experimental work performed, the major results, and final conclusions. This section should be written last.

3. **Introduction**: Concisely state the laboratory objectives and the importance of and application to the engineering community and to society.

4. **Procedure**: Clearly state how the experiment was performed, what equipment and supplies were used, what parameters were measured and how they were measured, and so forth. Diagrams or photos may be utilized to illustrate equipment and techniques that were utilized.

5. **Results and Discussion**: Results should include a minimum of one table, two figures (one sketch and one graph), and calculations for tensile strength. Discussions shall state the expected results, the achieved results, and the differences, if any, with reasons for the discrepancy, such as possible errors.

6. **Conclusions and Recommendations**: For example, comment on the relationship between width and tensile strength. Include recommendations for improvement of the experiment.

7. **Appendix**: Include the data sheet and sample calculations (including your spreadsheet) in a neat and clearly defined manner.

TABLE FORMAT:

A table such as the following will be created for each calculation of tensile strength:

TABLE 1 Tensile Test Measurements

Specimen Width (mm)	m (kg)	g (m/s^2)	W (N)	L$_1$ (mm)	L$_2$ (mm)	T (N)

Acknowledgment for this activity: Dr. Kenneth Leitch, as adapted from materials created by West Point Military Academy. See https://bridgecontest.org for more information.

Spot Speed Study Lab

OBJECTIVES:

1. Evaluate the general speed trends and posted speed limit on a roadway.

2. Process data into a useful graphical representation for engineering and informational purposes.

BACKGROUND:

1. Observing vehicle speeds is one way engineers measure safety on roadways. High speeds carry a high risk, whereas low speeds are relatively safe. In addition, observed speeds are used in capacity analysis, accident analysis, geometric studies, determining the need for pedestrian crossing, and before-and-after studies to assess the effectiveness of roadway modifications. Also, speed observations are compared to the posted speed limit to help determine whether the roadway is in need of greater law enforcement, realignment, or reconstruction. With so many important decisions being based on observed speeds, it is imperative that such speed data be collected correctly and presented effectively.

2. Many things affect the way drivers behave on the roadway. The proximity of roadside hazards, such as trees and utility poles; the frequency of driveways or curb cuts; and the presence of pedestrians affect how vehicle operators operate their vehicles. Since each driver is affected differently, the traffic engineer must look at a spectrum of vehicle-driver combinations, observe how each negotiates travel on the roadway, and define the range of speeds that may be expected at any particular time.

3. To perform a spot speed study, you must first select the location that is to be studied. The speed trap locations selected for study in this laboratory will be given by your instructor.

4. Groups will study either the eastbound or westbound directions on east-west roads and either northbound or southbound directions on north-south roads. In this study, general speed trends will be assessed. Therefore, data will be collected during daylight business hours, which is typical for this type of study.

5. Each group should be inconspicuous during data collection; that is, we don't want to look like we are gathering data about the drivers. This is important so that drivers do not alter their driving habits in the speed trap.

6. At each location, marks will be in place at a known distance. This information will be provided to you on the day of the lab. The other equipment required for this lab includes stopwatches and the attached data sheets.

PROCEDURE:

1. Form a group of two to three students.

2. Complete the information at the top of the data sheet prior to taking data.

3. After the top of the data sheet has been completed, you may begin to take speed data.

4. When the vehicle enters the speed trap, start the stop watch; when the vehicle leaves the speed trap, stop the watch; record the time in seconds on the data sheet.

5. Continue to take speed data until 100 records have been recorded or the end of the laboratory period has been reached.

DATA ANALYSIS:

1. Enter your collected data into an Excel spreadsheet.

2. Compute the speed for each data record. Speed in mi/h (mph) can be calculated using the following equation.

$$v = \frac{d}{1.467t}$$

Where d is the speed-trap length in feet and t is recorded time in seconds. (The constant 1.467 takes care of the conversion between ft/s and mph.)

3. Determine the minimum and maximum speeds recorded.

4. Develop speed groups, as shown in class. Your speed groups shall be in 2-mph increments. Therefore, if your lowest speed was 23.3 mph and your highest speed was 54.1 mph, you should have speed groups of 22–23.9, 24–25.9, 26–27.9, … 52–53.9, and 54–55.9 mph. Your instructor will also give you guidance on how to put together any plots required for this lab.

5. Calculate the frequency of each speed group. Create a histogram using the midvalues of the speed groups developed from your data.

6. Calculate the cumulative frequency for each speed group. Create a cumulative frequency curve using these data.

7. Determine the 50th-percentile speed (or median speed) and the 85th-percentile speed.

DELIVERABLES:

Students shall submit one report per group. This report shall consist of the following sections:

1. **Cover Page:** The report title page should contain the name of the lesson, date performed, class section and time, and all student group members' names.

2. **Abstract:** Clearly state in approximately two paragraphs a summary of the experimental work performed, the results, and your final conclusions. This section should be written last.

3. **Introduction**: Concisely state the laboratory objectives and the importance of and application to the engineering community and to society.

4. **Procedure:** Clearly state how the experiment was performed, what equipment and supplies were used, what parameters were measured and how they were measured, and so

forth. Diagrams or photos may be utilized to illustrate equipment and techniques that were utilized.

5. **Results and Discussion**: Include in your report your histogram and cumulative frequency curve. Provide a brief discussion of each in this section. Your discussion should include:

   a. How the 85th-percentile speed and posted speed limit compare.

   b. The percentage of traffic observed that was traveling over the speed limit.

   c. The distribution of speeds observed. Was the range of speed narrow or broad? Why?

   d. Potential sources of error.

6. **Conclusions and Recommendations**: For example, you may comment on the traffic that you observed. Did it look like many people were speeding? Did pedestrians influence traffic?

7. **Appendix**: Include the data sheet and sample calculations (including your spreadsheet) in a neat and clearly defined manner.

Data Sheet—Speed Trap Information

OBSERVERS:

Location: _______________________________________

Posted Speed Limit: _______________________________

Date: ____

Time Started: ____________________________________

Time Ended: _____________________________________

Weather: ____

Length of Trap: ___________________________________

Data Sheet—Time to Traverse Speed Trap

Vehicle No.	Time, s	Vehicle No.	Time, s	Vehicle No.	Time, s	Vehicle No.	Time, s
1		26		51		76	
2		27		52		77	
3		28		53		78	
4		29		54		79	
5		30		55		80	
6		31		56		81	
7		32		57		82	
8		33		58		83	
9		34		59		84	
10		35		60		85	
11		36		61		86	
12		37		62		87	
13		38		63		88	
14		39		64		89	
15		40		65		90	
16		41		66		91	
17		42		67		92	
18		43		68		93	
19		44		69		94	
20		45		70		95	
21		46		71		96	
22		47		72		97	
23		48		73		98	
24		49		74		99	
25		50		75		100	

Acknowledgment for this activity: Dr. Kenneth Leitch, as adapted from concepts presented in *Traffic & Highway Engineering* by Garber and Hoel, 5th ed., 2015.

Elastic Spring Lab

OBJECTIVES:

1. Verify the validity of Hooke's Law when applied to helical springs of various lengths and diameters.
2. Process data in tabular and graphical format to determine engineering quantities for design and analysis.

PROCEDURE:

1. Form a group of two to four students.
2. One report will be delivered by each group. Students in the same group will receive the same grade.
3. Each group will construct a mechanism to validate Hooke's Law for each type of spring to be investigated.
4. For each spring, students will increment the weights on the springs to collect data that relates the spring force to the amount of spring deflection (elongation).
5. Students can use predetermined weights, which are available in the lab. An example of predetermined weights would be to use 0.5 kg increments up to 5 kg on one spring, resulting in spring deformation values for each weight used.
6. A weighing scale can be used to determine if the stated value on the weight is indeed correct.
7. Students will create a table for the test measurements they conduct on each spring (see next page).
8. All results are to be in SI and USCS units, as indicated in the data form.
9. Students should submit a formal report on this activity.
10. For each spring, the following results are required:
    a. Table showing the spring test measurements.
    b. Figure that shows the spring force, F (lb_f or N) on the y-axis, as a function of spring elongation, ΔX (mm) on the x-axis.
    c. Calculation of the spring constant K (N/mm). Note that the slope of the trendline of your graph (with y-intercept set to zero) is the spring constant.

EQUATIONS TO BE USED:

Hooke's Law:

$$F = K \Delta X \tag{1}$$

where

    F = Spring force (lb_f, USCS) and (N, SI)
    K = Spring constant (lb_f/in, USCS) and (N/mm, SI)
    ΔX = Spring elongation (in, USCS) and (mm, SI)

Spring Force:

$$F = mg \tag{2}$$

where

    m = mass (kg)
    g = gravitational acceleration (= 9.81 m/s^2)

Conversions: 1 N = 0.22481 lb_f and 1 in = 25.4 mm

DELIVERABLES:

Students shall submit one report per group. This report shall consist of the following sections:

1. **Cover Page:** The report title page should contain the name of the lesson, date performed, class section and time, and all student group members' names.

2. **Abstract:** Clearly state in approximately two paragraphs a summary of the experimental work performed, the results, and your final conclusions. This section should be written last.

3. **Introduction:** Concisely state the laboratory objectives and the importance of and application to the engineering community and to society.

4. **Procedure:** Clearly state how the experiment was performed, what equipment and supplies were used, what parameters were measured and how they were measured, and so forth. Diagrams or photos may be utilized to illustrate equipment and techniques that were utilized.

5. **Results and Discussion:** Results should include a minimum of a table, two graphs (one USCS and one SI units), and calculations for spring constants for *each spring*.

   a. For each graph, create a linear trend line and determine its equation. Compare the slope of the trendline with the average spring constant value from your table. *Make sure to set the y-intercept of the trend line equal to zero.*

   b. Discussions shall state the expected results, the achieved results, and the differences, if any, with reasons for the discrepancy. Describe potential sources of error.

6. **Conclusions and Recommendations:** For example, you may comment on the springs and the test set up. Were the results useful?

7. **Appendix:** Include the data sheet and sample calculations (including your spreadsheet) in a neat and clearly defined manner.

TABLE FORMAT:

A table such as the ones that follow will be created for each spring (1, 2, 3, etc.) for this report.

TABLE 1 Test Measurements for Spring #1

Test#	Applied Force (lb$_f$)	Applied Mass (kg)	Applied Force (N)	Spring Elongation ΔX (in)	Spring Elongation ΔX (mm)	Spring Constant K (lb$_f$/in)	Spring Constant K (N/mm)
1							
2							
3							
4							
5							
6							
7							
8							
9							
10							
Avg	NA	NA	NA	NA	NA		

TABLE 2 Test Measurements for Spring #2

Test#	Applied Force (lb$_f$)	Applied Mass (kg)	Applied Force (N)	Spring Elongation ΔX (in)	Spring Elongation ΔX (mm)	Spring Constant K (lb$_f$/in)	Spring Constant K (N/mm)
1							
2							
3							
4							
5							
6							
7							
8							
9							
10							
Avg	NA	NA	NA	NA	NA		

TABLE 3 Test Measurements for Spring #3

Test#	Applied Force (lb$_f$)	Applied Mass (kg)	Applied Force (N)	Spring Elongation ΔX (in)	Spring Elongation ΔX (mm)	Spring Constant K (lb$_f$/in)	Spring Constant K (N/mm)
1							
2							
3							
4							
5							
6							
7							
8							
9							
10							
Avg	NA	NA	NA	NA	NA		

Acknowledgment for this activity: Dr. Kenneth Leitch.

Spreadsheet Lab

OBJECTIVES:

1. Utilize Microsoft Excel to determine the standard deviation, mean, and median of a data set(s).
2. Make predictions based on a data set(s).

PROCEDURE:

1. Download the data set(s) supplied by your instructor.
2. One memo report will be delivered by each student. Utilize Microsoft Excel or equivalent to process the data.

DELIVERABLES:

1. Summary memo (one per student) addressing the following for the data set(s):

   a. Standard deviation

   b. Mean

   c. Median

   d. Any probabilities or predicted values required by the class instructor, which include the following:

      i. ____

      ii. ____

      iii. ____

      iv. ____

2. Memo attachments

   a. Spreadsheet

   b. Graph

Acknowledgment for this activity: Dr. Erick Butler.

Launcher Project

OBJECTIVES:

1. Design, construct, and test a launcher.

2. Collaborate as a team to meet the required project parameters.

3. Produce a comprehensive report that documents the design, construction, and testing of the launcher, including written documentation, illustrative sketches, and test data.

PROCEDURE:

1. Form a group of two to three students.

2. Groups are tasked to design, build, and test a launcher within these specifications:

   a. The launcher can use any launch system (catapult, trebuchet, spring/elastic, hydraulic, compressed air, etc.) **EXCEPT** combustion-based systems (combustible fuel, gun powder, rockets, potato guns, etc.).

   b. The launcher must be a repeatable system (i.e., capable of multiple launches).

   c. The launcher must be adjustable to launch a payload at a variety of angles. This angle must be measureable and accurate at the time of launch. A protractor (or similar device) is helpful for measuring this angle.

   d. The launcher must be portable and mobile, meaning that the members of the group should be able to carry it by hand and set it up in 5–10 minutes (i.e., no trailer-mounted launchers).

   e. No external means of energy is allowed aside from that provided by the human operators (i.e., no electricity, motors, etc.). *Note:* Compressed air is allowed, provided the air is compressed solely by human power.

   f. Use of other equipment and supplies is not restricted, but equipment and supplies purchased by the university are restricted to a set budget per group.

DELIVERABLES:

1. Materials List: Write a memo listing the requested items. Preference will be given to materials that have descriptions and prices from Ace Hardware and Walmart in Canyon, Texas.

2. Initial design, with consideration to originality, planning, ease of use, implementation, and conformity to design specifications.

3. Performance, with consideration to both distance and consistency.

4. Efficiency of design, including performance vs. weight and ease of use.

5. Project report with documentation of design, construction, and testing.

6. Bonus (optional):

   - Most accurate

- Lowest cost
- Aesthetics (i.e., nicest looking)

BACKGROUND INFORMATION: LAUNCHER KINEMATICS

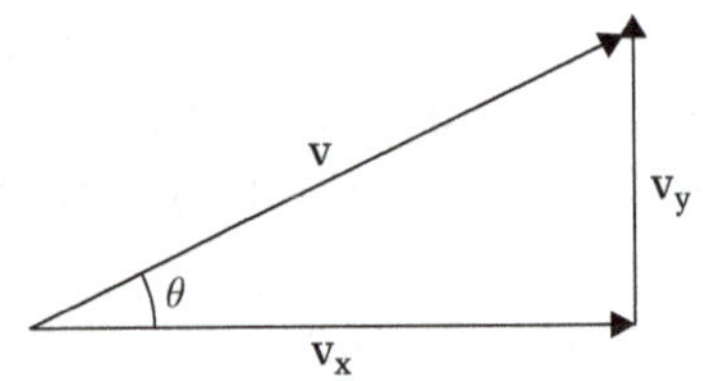

Rectilinear projectile motion

v = initial velocity

$v_x = v \cos \theta$

$v_y = v \sin \theta$

$$x = v_x t \quad (1)$$
$$y = v_y t - 0.5a\, t^2 \quad (2)$$

where in this case, $a = g = 9.81$ m/s^2 or 32.2 ft/s^2.

By measuring the distance the ball travels (x) and knowing that y = 0 when the payload hits the ground, there are now two equations ((1) and (2)) which can be solved for the two unknowns (t and v).

The initial launch velocity and flight time should be calculated for every launch. Average values and standard deviation can be calculated for each launch angle to determine the consistency of the launcher. Vary the launch angle to find the optimum launch angle.

Ohm's Law and Circuits Lab

OBJECTIVE:

Verify Ohm's law by placing voltage across a resistor (or resistors) and measuring the current. Different resistor values will be used, and voltage will be measured across each resistor and compared to the results of Ohm's law. We will determine errors.

EQUIPMENT:

- Power supply with variable voltage output. The power supply contains an internal voltmeter.
- Multimeter or Volt-Ohm meter that has a current measurement down to the 500 μA range. A Simpson meter may be used for the ammeter.
- Five resistors from 500 Ω to 1 MΩ.
- Banana leads with alligator connectors on the end (one for power supply and one for the meter).

PRECALCULATIONS:

Assuming a maximum of 50 V, calculate the current through each of the resistors. Use Ohm's law $I = V/R$. Also calculate the power for the resistor: $P = V^2/R$ or $P = I{\cdot}V$.

   If the power is calculated to be higher than about 1 Watt, do not test the resistor at that power. Calculate power again. Notice that if you cut the voltage in half, the power goes down by a factor of four (using the V^2/R equation).

Resistor Value (Ohms)	Current at 50V	Power at 50V	Current at 30V	Power at 30V	Current at 10V	Power at 10V

PROCEDURE:

- Select 5 resistors and measure their resistances with an ohmmeter. Record their values. On most scales that you will use (2K, 20K, 200K), the reading is in kilo-ohms. So 0.5 would be 0.5 kΩ, 3.3 would be 3.3 kΩ, 47 would be 47 kΩ, as long it is set on a scale that is _____K.

- Connect the volt-ohmmeter to a set of alligator clip leads. Connect the black lead into the – (or common) connector in the meter. Connect the red lead into the + (or it may be labeled "Ω" on the meter).
- Connect the alligator clip leads to the power supply. Make sure the red lead is in the positive output banana connector. We will always use this convention, red for + and black for –. We will also always use – for grounded systems.
- Clip the black clip from the power supply to one side of a resistor.
- Clip the red lead of the ammeter to the red (+) side of the power supply.
- Clip the black lead of the ammeter to the other side of the resistor.
- Determine the estimated current for that resistor (from the precalculations). Set the meter on the Amp scale and set it to the next range up from the estimated current for that resistor (e.g., if the estimated current is 150 mA, set it on the 200 mA range).
- Draw a diagram of your setup.
- STOP! DO NOT PROCEED! Have the lab instructor verify your work before you proceed.
- After the lab instructor has verified your setup, turn on the power supply.

  - First turn the current knobs (on top) clockwise all the way.
  - Turn the right (bottom) knob very slowly until the voltage reaches just below 10 V. Then turn the voltage (bottom knobs) up to 10 V and measure the current.
  - Use the left lower knob to fine-adjust the voltage on up to 10 V. If you need to, turn-adjust the range of the meter so that the reading is within range of the current.

- Record the current for that voltage and repeat for 20 V, 30 V, 40 V, and 50 V.
- On the table below, record the resistor value and the current for each voltage.

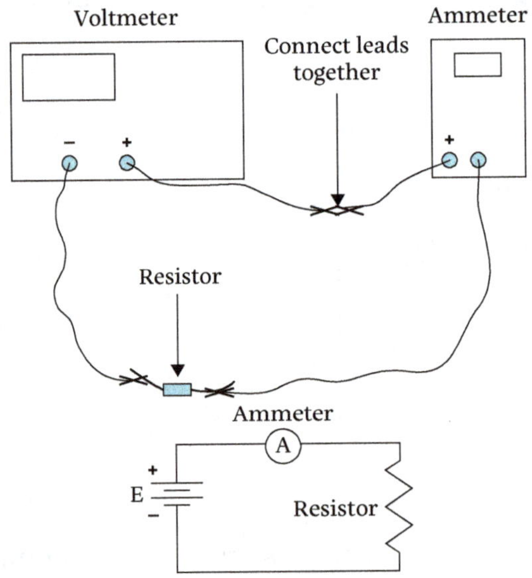

Resistor Value (Ohms)	Current at 10 V (mA)	Current at 20 V (mA)	Current at 30 V (mA)	Current at 40 V (mA)	Current at 50 V (mA)

DELIVERABLES:

1. You should have a data table with about 5 voltage/current values for each resistor.

   a. Using Ohm's law, determine what the current should be through each resistor

   b. Compare that with the actual readings that you measured (you did a few of these in the precalculations).

2. Determine the % difference you get. Do this for one of the current readings for each resistor. Recall that **% error = 100 x (actual reading – calculated reading) / (calculated reading)**

3. On one graph, plot the current vs. voltage for two of the resistors. Pick the two resistors that are closest to each other in value for the plot.

4. Write up any conclusions you might have. How does the plot you made characterize Ohm's law? What is the difference between the two plots you made? What does the slope indicate? If there is an error, what instrument introduced the most error? The voltmeter (in the power supply), the ammeter, or the ohmmeter?

Acknowledgment for this activity: Michael Blasdel.

Water Hardness Lab

OBJECTIVES:

1. Introduce students to an environmental engineering topic.

2. Perform an experiment to simulate a water softening technique.

BACKGROUND:

1. Water hardness is the measurement of calcium (Ca^{+2}) and magnesium (Mg^{+2}) cations found in water. High amounts of these cations equates to water being hard. Low amounts of these cations produces soft water.

2. The effects of hard water can be seen whenever water leaves a white substance on bathroom fixtures, sinks, and toilets. This white substance is commonly known as scum. Hard water can also damage pipes by creating a buildup of calcium deposits inside the pipe. Municipalities that experience hard water are usually those that primarily use groundwater as their water source.

3. Water hardness is most commonly measured as a concentration of milligrams per liter as calcium carbonate (mg/L as $CaCO_3$).

4. There are two categories of hardness—carbonate hardness and non-carbonate hardness. Carbonate hardness is the combination of calcium and magnesium cations with the bicarbonate ion (HCO_3^-). Non-carbonate hardness combines the remaining calcium and magnesium with non-bicarbonate anions sulfate (SO_4^{2-}) and chloride (Cl^-).

5. Carbonate hardness is of most concern, and it is acceptable to have non-carbonate hardness in drinking water. This is very apparent in bottled drinking water. Because many bottled drinking water companies use technologies that can remove almost all ions from water, non-carbonate hardness compounds such as magnesium sulfate are added back in for taste purposes.

6. Environmental engineers are involved in designing treatment methods to remove the calcium and magnesium ions from hard water. This process is known as softening.

7. Water softening adds the chemicals lime (calcium hydroxide, $CaOH$) and soda ash (sodium carbonate, Na_2CO_3) for the purpose of forming solids (precipitates) consisting of either calcium carbonate ($CaCO_3$) or magnesium hydroxide ($MgOH$). These solids are then separated from the water, making it softer. Softening requires an understanding of chemical reactions and pH adjustments. These chemical reactions determine the proper amount of lime and soda ash needed to precipitate out calcium carbonate and magnesium hydroxide.

8. While many of the industrial chemicals used are not easily accessible to the common public, there is a very simple method for softening water that can be performed at home. In this lab, students will use a common household chemical to attempt to reduce the calcium and magnesium ions within the water.

MATERIALS:

washing soda

aquarium test strips

mixing container

disposable foam cups

empty plastic water bottles

coffee filters

plastic spoon

timer

disposable gloves (**Note: Wear gloves while handling chemicals.**)

scale

PROCEDURE:

1. Get into groups of between three and four students.

2. Fill the container with water from the sink with about 24 ounces of water.

3. Take a filter strip and dip it into the water for about 1 second. Shake off any excess water from the strip. Take out the strip and wait about 60 seconds. Read and record the color, the closest hardness concentration, total alkalinity (carbonate hardness), and pH on the worksheet. The results should be determined by comparing the strip to the freshwater comparison chart.

4. Calculate the non-carbonate hardness using the non-carbonate hardness equation. Record the value on the lab handout.

5. Take the disposal foam cup and put it on a scale. Zero the scale.

6. Pour washing soda into the foam cup until about a quarter of the cup is filled. Observe the weight of the soda (in milligrams). Record on the data sheet. Transfer the washing soda from the cup into the mixing container filled with water.

7. Vigorously stir the water with a plastic spoon for two minutes. This will allow the washing soda to dissolve into the water.

8. Stir very slowly for five more minutes.

9. At the end of the stirring, let the mixture settle for 30 minutes. While allowing the mixture to settle, take one of the plastic bottles and cut the top off. The top piece should form a cone.

10. Take two coffee filters and place them inside the top part of the water bottle. The lip of the bottle should be face down. Place directly on top of the other empty water bottle.

11. Wrap tape around the two pieces. Make sure the two pieces fit securely. This will form a filter.

12. While waiting for the mixture to settle, complete Discussion Questions 1–4. Answer on the lab handout.

13. After settling, take the mixture from the container and pour into the filter. The purpose is to filter out developed solids created during the softening process. Prepare enough solution so that a filter strip can be immersed in the water.

14. Take off the tape and pour the filtered water into another foam cup.

15. Take a filter strip and dip it into the water for about 1 second. Shake off any excess water from the strip. Take out the strip and wait about 60 seconds. Read and record the color and the closest hardness concentration and pH on the worksheet. The results should be determined by comparing the strip to the pH strip container or chart provided with the pH strips.

16. Calculate the percent change in hardness. Record on the lab handout.

Percent change = [initial hardness (mg/L)] – [final hardness (mg/L)]/[initial hardness (mg/L)]) × 100

DISCUSSION QUESTIONS:

1. What are some activities done in the home that are affected by water hardness? Have you ever experienced hard or soft water? What did it feel like?

2. Which type of water do you prefer? Bottled or tap water? Why?

3. If the average person drinks 64 oz per day and it costs $0.0056/gal for tap water and $9.46/gal for bottled water, how much would it cost a family of 4 to drink tap versus bottled water in one year? Based on this question, if you still drink bottled water, would you ever convert? Why or why not? (Note: 1 oz = 0.0078125 gallons)

4. If drinking water is safe coming out a water treatment plant (WTP), what would be some causes of water tasting funny when you drink it out of the tap?

5. Carbonate hardness is the primary focus for environmental engineers. However, washing soda contains sodium carbonate (soda ash); therefore, this experiment most likely reduced the presence of non-carbonate hardness. The removal of non-carbonate hardness from calcium is expressed by the following equation:

$$Ca^{+2} + [2Cl^- \text{ or } SO_4^{2-}] + Na_2CO_3 \rightarrow CaCO_3 + 2Na^+ + [2Cl^- \text{ or } SO_4^{2-}]$$

(calcium ions) (chlorine or sulfate ions) (sodium carbonate) (calcium carbonate) (sodium ions) (chlorine or sulfate ions)

Using the non-carbonate hardness value calculated before treatment, determine the amount of soda ash needed to reduce the non-carbonate hardness by half. (Hint: Use the soda ash required equation.)

6. Was the pH after treatment greater than 8.4? What would cause the pH to increase?

Acknowledgment for this activity: Dr. Erick Butler.

Figure Credits

Examples of Technical Writing for Engineers

In this section, examples of memoranda, technical research papers, and full lab/project reports are given as guidance for students. These are by no means the most definitive examples, nor are they perfect examples of written English, but they can help students in their quest to write more effectively.

Students and practicing engineers are often amazed by the amount of writing they will have to do. Their writing, thankfully, is factual (instead of fictional) and presents the results of experimentation, analyses, and designs to their intended audience in a succinct, direct manner.

It is advisable to consult English style guides and to use spelling and grammar checking tools built into word processing and other computer software. However, realize that these tools have their limits and that the author(s) should always check his or her (or their) work to make sure that it is factual, correctly calculated, and polished in its appearance and, most of all, that it makes sense when read or spoken aloud.

Helpful points in regard to technical writing for engineers:

- Writing should be objective. Do not use words such as *I*, *we*, *us*, *you*, *they*, *them*, etc., since they are not in an objective view.
- Use proper English, avoiding slang and contractions.
- If special engineering terms are used, describe them in brief, depending on the audience.

  - Example: Engineering *strain* refers to the change in length per unit length of an object and is often noted in terms of percentage.
  - Example: The *modulus of elasticity* is a measure of the ability of a material to deform under loading, such that it will return to its original shape when the loading is removed. It has typical units of force per unit area such as lb/in^2 (psi) or N/m^2 (Pa).

- Use appropriate units with numerical values, as described in this text.
- Superscripts, subscripts, Greek symbols, and other special symbols and formatting are important. Learn to use these in word processors, spreadsheets, presentations, and other forms of technical communication.
- Figures, images, and tables should be labeled and clear.
- Include references, where appropriate. Google Scholar (https://scholar.google.com) is a great place to find peer-reviewed technical articles and other information to use as appropriate references.
- When in doubt, ask the professor or instructor in the workplace or a reputable coworker or superior on what is expected in the technical document or presentation that you are producing.

Sample Memoranda

See the following pages for sample memoranda.

MEMORANDUM

To:	Dr. Kenneth R. Leitch, PE
From:	Russ T. Steele
Date:	02 Jun 2016
Re:	ENGR 1171, Engineering Ethics, Review of Erin Brockovich Movie

The film *Erin Brockovich* (2000) is the true story of a huge unethical scandal that was discovered during her work with a lawyer's firm. The film shows the trials and tribulations of a single mother and legal clerk who blows the whistle on energy giant Pacific Gas and Electric Company and wins. Like Enron Corp., the company was exposed for unethical behavior and hundreds of millions were paid back to the small California desert community that was damaged due to chemical discharges to the environment.

The film opens as an unemployed single mother struggling to make ends meet with her three children. Her legal clerk journey begins by being a victim in a traffic accident; she decides to sue the other driver. After a courtroom loss due to her actions during the proceedings Brockovich finds herself in a tight spot knowing she needs to provide for her family. Erin Brockovich, having little education, decides to take matters into her own hands and gets herself a second chance being a legal clerk for the lawyer who represented her for the trial. The lawyer is hesitant in his choice but is very intrigued by the tenacity of the single mother and does feel guilty for losing her case.

As Brockovich's career begins, she stands out as a legal clerk who is not afraid to go against the grain to get her point across. She is tasked with reviewing real-estate files as part of her workload. She then begins to find irregularities in the documents finding what seems to be irrelevant health files in a real-estate deal, Brockovich then begins to do some digging of her own. While juggling her children, new boyfriend, and new job, she goes to visit the community where Pacific Gas and Electric is trying to buy up all the land. Once there, she begins to find the true reason why the health records were tied in with the real-estate files.

The main woman of interest to Brockovich was Donna Jensen who was the occupant of the house that Pacific Gas and Electric was interested in buying. She was astounded to find that the family has been plagued with diseases and have been receiving medical attention at PG&E's expense. Having this fact she decides to dig deeper into why this energy giant would be paying for the medical bills of this family and why they are interested in buying their land. The reasoning that PG&E gives for their donation to the Jensen family was because of the hexavalent chromium in the water. She then begins to look into how chromium is related to the diseases plaguing everyone using the ground water in the area.

Brockovich discovers that the groundwater is extremely contaminated with carcinogenic hexavalent chromium due to PG&E's poor disposal process. PG&E has been telling the community that is was a safe form of chromium, which seems to be a lie. As the film advances Brockovich begins to dig deeper into the residents of Hinkley to find that most of the people have had some medical problems. All the medical problems were taken care of by PG&E and their doctors. She then decides to bring PG&E to justice for what they have done; with the help of her boss they bring up a class action lawsuit.

PG&E headquarters then states they had no idea of the matter and tries to shift the blame to the Hinkley Plant itself, which would decrease the payout substantially. As this case gets closer to court they find out that they can be more effective if they get a majority of the residents of Hinkley affected by the chromium to go along with a binding arbitration. While Brockovich is completing this task, an employee of PG&E who was in charge of destroying all documents relating to PG&E and the chromium confronts her. It turns out he kept a memo linking the headquarters and Hinkley Plant in regard to the dangerous chromium that was in the groundwater. This was the nail in the coffin for PG&E and in turn they had to pay out $333 million to all the residents of Hinkley who were affected by the groundwater catastrophe.

In the end Brockovich gets a sizable check showing the gratitude of what she had done for the company, which was in the amount of $2 million dollars. Overall this was a great film showing that anyone can blow the whistle on unethical behavior.

MEMORANDUM

To:	Dr. Kenneth R. Leitch, PE
From:	John P. Smith, Kimberly C. Jones, Omar Perez
Date:	14 Mar 2011
Subject:	Tensile Creep Lab Results

A laboratory was performed to display the creep property of materials; creep is able to occur due to sustained tensile loading and gaps of materials at the atomic level. To observe the creep phenomenon, a lead specimen was placed under a constant load of 450 N at room temperature. This material was a prime candidate for the experiment due to its high creep rate and its soft nature. The reason this is possible and the reason lead has a high creep rate is because the observed room temperature 294 K (21°C) is approximately half of the material's melting temperature of 601 K (328°C).

After approximately 56 minutes of a constant load of 450 N, the lead specimen broke due to the tensile creep process. Measurements were taken at two-minute intervals to track the change in elongation over time, which was used for calculation tablet and provided plot. Strain is calculated as:

$$(\text{Strain})\,\varepsilon = \frac{l - l_0}{l_0}$$

For this lab's lead specimen, a creep rate of 0.0019 m/m/s was determined. Because the break occurred near the beginning of the third phase of the creep process, it is difficult to see the transition into the third phase. Note that three phases total exist in the creep process—a rapidly changing first phase (extremely short), a linear second phase (longest section), and a short third phase with larger change than phase two, which leads to breaking.

Attachments:

- Strain vs. Time Plot
- Calculation Table

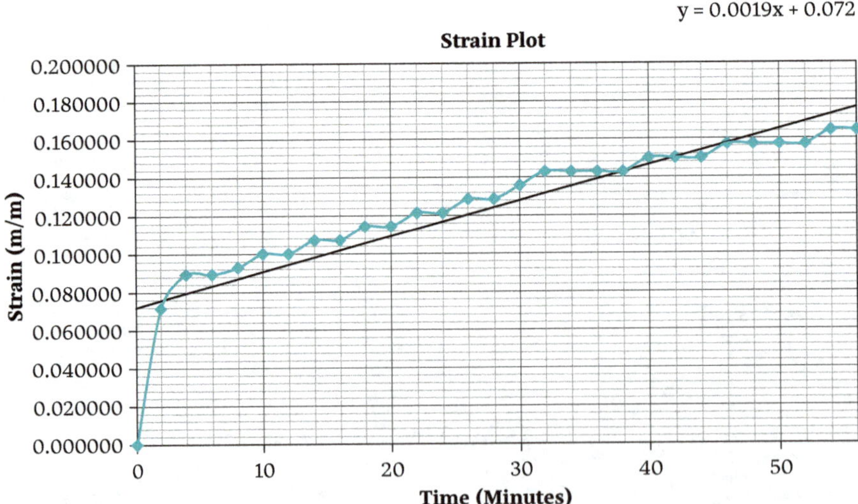

FIGURE B.1 Strain vs. Time Plot

TABLE 1 Length and Strain vs. Time of Lead Specimen

Time (Minutes)	Length (m)	Stress (MPa)	Time (Minutes)	Strain (m/m)
0	0.1400	11.22	0	0.000000
2	0.1500	11.22	2	0.071429
4	0.1525	11.22	4	0.089286
6	0.1525	11.22	6	0.089286
8	0.1530	11.22	8	0.092857
10	0.1540	11.22	10	0.100000
12	0.1540	11.22	12	0.100000
14	0.1550	11.22	14	0.107143
16	0.1550	11.22	16	0.107143
18	0.1560	11.22	18	0.114286
20	0.1560	11.22	20	0.114286
22	0.1570	11.22	22	0.121429
24	0.1570	11.22	24	0.121429
26	0.1580	11.22	26	0.128571
28	0.1580	11.22	28	0.128571
30	0.1590	11.22	30	0.135714
32	0.1600	11.22	32	0.142857
34	0.1600	11.22	34	0.142857
36	0.1600	11.22	36	0.142857
38	0.1600	11.22	38	0.142857
40	0.1610	11.22	40	0.150000
42	0.1610	11.22	42	0.150000
44	0.1610	11.22	44	0.150000
46	0.1620	11.22	46	0.157143
48	0.1620	11.22	48	0.157143
50	0.1620	11.22	50	0.157143
52	0.1620	11.22	52	0.157143
54	0.1630	11.22	54	0.164286
56	0.1630	11.22	56	0.164286

Sample Report in APA Format

See the following pages for an example report in APA format.

Enron's Ethical Duties to Outside Stakeholders

Jane A. Good-Student

Awesome State University

Engineering Ethics

ENGR 1171

Instructor: Dr. Kenneth R. Leitch, PE

05 Jun 2015

ABSTRACT

In this report, the author will discuss how Enron Corporation failed in its ethical duties to its stakeholders: the shareholders, employees, and general public. The author discusses how various ethical theories (Kantian, utilitarian, and virtue) can be applied to the Enron case. Enron Corporation collapsed in 2001 amid a financial accounting scandal perpetrated by its top leadership. The gross ethical misbehavior of the company's leadership and lack of due diligence of the auditing firm of Arthur Andersen resulted in the fall of both corporations and influenced the passage of the Sarbanes-Oxley Act of 2002.

BACKGROUND INFORMATION

Before looking at the ethical issues in the collapse of Enron, it is helpful to examine the human toll that the scandal has exacted. Fifteen high-level officials of Enron were prosecuted, including its top three executives. Enron CEO Jeffrey Skilling was sentenced to 24 years in prison (Hays, 2007). Financial chief Andrew Fastow pled guilty to his major role in disguising company losses in a series of shell companies, where approximately 2/3 of the corporate losses were hidden, ending up with a 10-year prison term (Frey, 2004). Chairman Kenneth Lay was convicted on ten counts, which carried a prison term of approximately 20 to 30 years, but died in 2006 before sentencing. Enron Corporation laid off 4,000 employees as a result of its bankruptcy and dissolution. Enron's auditor, Arthur Andersen, dismissed 28,000 staff as a result of the surrender of its public auditing license. As a result of Enron, Arthur Andersen, WorldCom, and other high profile financial scandals, the US Congress passed the Sarbanes-Oxley Act of 2002 to tighten oversight of financial reporting for publicly held corporations.

A QUESTION OF ETHICAL BEHAVIOR

A hallmark of Kantian ethical theory is to do no harm to others. Enron Corporation did have a code of ethics, but it appears to have been widely disregarded by top management. A code of ethics is ineffective if it is not enforced and cultivated by corporate leadership. Enron began life as sleepy Houston Natural Gas, a steady business that delivered a useful, if not exciting, product to homes and businesses. In the 1990s, the company began to expand into new business ventures such as energy trading (related to the original business) and trading of other commodities, such as broadband internet (far removed from the natural gas transmission business). It is estimated that before the company failed, it was trading up to 1,800 different types of commodities. The company grew rapidly, bringing in a large number of new employees whose style clashed with those who worked in the traditional natural gas transmission business. Leadership as expressed by Boa et. al (2007) entails commitment to developing employees in ethical and decision-making skills. This was arguably not the case at Enron, as young, aggressive executives such as Skilling and Fastow moved into the leadership of the company, leaving the experienced Lay to be a traveling ambassador that extolled the Enron Way of making money.

It is in this backdrop of a corporation run by daredevil, thrill-seeking executives that Enron began to falter. An elaborate network of shell corporations were used by Andrew Fastow to place the majority of Enron's losses from failed ventures. Long-time auditor Arthur Andersen precipitated the move toward bankruptcy with restatement of Enron's financial position, showing that the company was *losing* instead of *earning* money. It is purported that Arthur Andersen knew of financial irregularities since 1998, nearly four years before the bankruptcy filing in December 2001.

As previously mentioned, leadership of Enron and its auditor Arthur Andersen were not following the fundamental imperative of Kantian ethics. It can also be argued that while management might have been trying to follow a utilitarian ethical view in that they attempted to do *good* for a large number of people who worked for and invested in Enron by attempting to keep the stock value high, they ultimately fail this ethical test, as they perpetrated large scale damage to themselves legally and financially as well as by causing a disservice to the thousands of Enron and Arthur Andersen employees that lost their jobs, to stockholders of both corporations, to suppliers, to customers, and to the public. In short, Enron fails to "promote human welfare by minimizing harms and maximizing benefits" (Beauchamp, 2009).

WHISTLEBLOWING: AN ETHICAL IMPERATIVE

Sherron Watkins was the Enron Vice President that sent a letter to Chairman Kenneth Lay telling of her concern that the company was about to "implode in a wave of accounting scandals" and that there was still time to save the company (Analysis: Enron, 2002). At first glance, it seems ironic that Ms. Watkins would sound an alarm, as she previously worked at Arthur Andersen auditing Enron and then joined Enron in 1993 to work for Andrew Fastow in finance. She was vested in the system, but knew that her 18 years at Arthur Andersen and Enron would be worthless if (or when) Enron were to "implode." Watkins is considered a whistleblower due to the internal letter to Lay that was made public around the same time in February 2002 (three months after Enron's bankruptcy filing) that she appeared in front of the Senate Commerce Committee to testify about her fellow Enron executives. Her testimony is in line with virtue ethics in that she developed a virtuous character from skills developed as a certified accountant and as an executive. While her fellow executives dodged questions from politicians and courts, she spoke openly about what she observed in the corporate culture at Enron. Her example has not gone unnoticed, as she was named one of Time magazine's persons of the year 2002. In Watkins' own words, "I am privileged to be in demand as a lecturer, speaking to business leadership conferences, graduate business schools, and religious groups, using Enron as an example of failed leadership. But even still, I face criticism about whether my actions were enough, or worse, whether my actions were motivated to advance myself somehow. This seems to be the norm for people who take any form of whistleblower action. When we, as a society, pick apart the whistle-blowers, we actually discourage others from taking a stand in the future." Her response is consistent with virtue ethics, as described by Greek philosopher Aristotle; such virtuous character must be cultivated in an individual (Beauchamp, 2009).

CONCLUSION

A corporation needs to ensure that it has a code of ethics that is actively promoted by its leadership. Corporations need to be especially vigilant when they grow quickly to ensure that an ethical culture is nurtured. The study of ethics and ethical theories would be of great use to corporate executives and employees to help foster the growth of an ethical culture. The reforms of the Sarbanes-Oxley Act of 2002 are a good start, but ultimately, it is up to individuals and corporations to maintain propriety and a high ethical standard in all of their business dealings.

REFERENCES

Beauchamp, T., Bowie, N., Arnold, D. (2009). *Ethical theory and business* (8th ed.). Upper Saddle River, New Jersey. Pearson Prentice Hall.

Boa, K., Buzzell, S., Perkins, B. (2007). *Handbook to leadership: Leadership in the image of God*. Atlanta, Georgia. Trinity House Publishers.

Edwards, B. (24 Feb 2002). Analysis: Enron whistle-blower Sherron Watkins testifies before the Senate Commerce Committee, *NPR*.

Frey, J. (15 Jan 2004). The crash of a symbol of Enron greed. *Washington Post*, p. C01.

Hays, K. (19 Jun 2007). Ex-Enron insider gets 27-month term: Ex-broadband CEO is last of 15 pleading guilty to be punished. *Houston Chronicle*.

Koerwer, V. (22 Jun 2004). Featured interview Sherron Watkins, former vice president for corporate development of Enron. Accessed 21 Feb 2009, from http://www.allbusiness.com/educational-services/business-schools-computer/290768-1.html

Sample Full Lab and Project Report

See the following pages for an example report, with instructor comments included in square brackets and noted with instructor initials "NGA."

Fluid Mechanics Lab #2
Flow Rate

Prepared by: Johnny B. Goode

Group members: Johnny B. Goode, LeeAnne Wright, and Chuck Roast

For: Dr. N. Gene Ayre

February 20, 2014

ABSTRACT

The Flow Rate Lab examined the methods used to calculate Volumetric Fluid Flow both experimentally and theoretically. By having known values for discharged water weight, time of discharge, and density it is possible to calculate both mass and volume flow rates directly. With known values for fluid initial height and fluid exit area it is possible to theoretically calculate volume flow rate using degenerate versions of the Energy Equation known as Bernoulli's Equation and Torricelli's Equation. The experimental value for volume flow rate was significant lower than the theoretical value. The search for the cause of the error uncovered the limitations of Bernoulli's Equation and the importance of friction and head loss in calculations.

[Nice abstract 10/10 points]

TABLE OF CONTENTS

LIST OF FIGURES

LIST OF TABLES

NOMENCLATURE

Notation	Description	Units
P	Pressure	Pa
V	Velocity	m/s
$\dot{V}$	Volumetric Flow	m³/s
$\dot{m}$	Mass flow Rate	kg/s
A	Area	m²
D	Diameter	m
z	Height	m
g	Gravity	m/s²
ρ	Density	kg/m³
π	Pi (constant)	N/A
P_{atm}	Atmospheric Pressure	Pa

Format 10/10

INTRODUCTION

The purpose of the Fluid Flow lab was to experimentally determine the volume flow rate of water under two different conditions. The two conditions were out of a spigot at the bottom of a reservoir (5 gallon bucket), and through a siphon tube from the same reservoir. The volume flow rate is the volume of fluid flowing through a cross section per unit time [1]. This lab experiment investigates the use of Bernoulli's Equation in calculating theoretical volume flow rate while assuming no flow loss. These values can then be compared to the flow rate values found through experimentation.

By comparing the flow rate values, students can get a better understanding of Bernoulli's Equation, its limitations, and its relation to the Energy Equation, which is more general and accounts for flow loss.

[Well done 10/10 points]

THEORY

To calculate the volume flow rate of water theoretically using Bernoulli's Equation, and directly through experimental data, it is required to measure the time it takes the water to drain approximately 0.5 gallons and also take some other measurements. These measurements include the height in the reservoir bucket before flow, the weight of the discharged water in the receptacle bucket, spigot inner diameter, dry mass of receptacle bucket, and siphon tube inner diameter, and the length of the siphon tube.

Bernoulli's Equation for steady, incompressible flow is as follows [1]:

$$\frac{P_1}{\rho} + \frac{V_1^2}{2} + gz_1 = \frac{P_2}{\rho} + \frac{V_2^2}{2} + gz_2 \tag{1}$$

Bernoulli's Equation is then put in terms of head by dividing through by gravity (g) to obtain:

$$\frac{P_1}{\rho g} + \frac{V_1^2}{2g} + z_1 = \frac{P_2}{\rho g} + \frac{V_2^2}{2g} + z_2 \tag{2}$$

In the theoretical calculations for volume flow rate the following assumptions were made: $P_1 = P_2 = P_{atm}$, $V_1^2 \ll V_2^2$ therefore V_1 is approximated to zero compared to V_2,

z_2 = Reference and thus is defined as zero. [Excellent. Listing assumptions.]

Based on these assumptions Bernoulli's Equation (2) reduces to:

$$z_1 = \frac{V_2^2}{2g} \tag{3}$$

Rearranging the equation to solve for V_2 yields:

$$V_2 = \sqrt{2gz_1} \tag{4}$$

This is known as Torricelli's Equation [1].

Volumetric flow rate ($\dot{V}$) is defined as:

$$\dot{V} = V_2 A \tag{5}$$

After substituting in Equation 4 for V_2, and the equation for area, Equation 5 becomes:

$$\dot{V} = \frac{\pi}{4} D^2 \sqrt{2gz_1} \tag{6}$$

The theoretical value for volume flow rate can then be calculated using Equation 6. Calculating volume flow rate using experimental data is accomplished using the equation:

$$\dot{V} = \frac{\dot{m}}{\rho} \tag{9}$$

With $\rho = 1000 \ \text{kg/m}^3$, and equal to weight of the discharged water divided by measured time.

[The theory section clearly develops the theoretical equation and lists the assumptions used in solving or simplifying the equations. It also provides the basis for the experimental data to which the theory will be compared. 15/15 points]

MATERIALS AND EQUIPMENT

<u>Flow rate lab required equipment</u>:

5+ gallon bucket with valve controlled spigot

Receptacle bucket

Clear plastic siphoning tube

Timer

Vernier Caliper (mm)

Yardstick (in)

Digital scale (g)

Tap water

EXPERIMENTAL PROCEDURE

A. <u>Measurement 1 (Spigot)</u>

1. Obtain 5+ gallon bucket with valve-operated spigot and after making sure valve is closed, fill the bucket with tap water to at least the 5-gallon mark.

2. Measure the height of the initial water level (z_1) using the yardstick.

3. Measure inner diameter (D) of the end of spigot using Vernier caliper.

4. Measure the dry weight of the receptacle bucket using digital scale.

5. Place the filled 5+ gallon bucket on a table or other raised surface.

6. Place the receptacle bucket under the spigot.

7. Open the valve and allow approximately 0.5 gallons to drain into the receptacle bucket, while using timer to measure time needed to drain the approximate half-gallon.

   a. Make sure the spigot control valve is fully opened each time the water is drained.

   b. Measure the weight of the receptacle bucket with drained water using the digital scale.

   c. Measure the height of the water (z_1) in the 5+ gallon reservoir bucket after each half-gallon increment.

   d. Record drain time, discharged water weight, and remaining reservoir water height after each run using yardstick.

8. Repeat step seven around 7–8 times.

B. <u>Measurement 2 (Siphon)</u>

1. Refill reservoir bucket to at least the 5-gallon mark.

2. Measure the initial water level (height) using the yardstick.

3. Place the filled 5+ gallon bucket on a table or other raised surface.

4. Measure the inner diameter of the siphon tube using Vernier caliper.

5. Place the receptacle bucket under and in front of the reservoir bucket.

6. Using the siphon, drain water from the 5+ gallon reservoir to the receptacle bucket in approximately one-half-gallon increments while using the timer to measure time needed to drain the approximate half-gallon.

   a. Make sure there are no air bubbles in siphon tube and that the flow remains steady throughout draining.

   b. Measure the weight of the receptacle bucket with drained water after each half gallon increment using digital scale

   c. Measure the height of the remaining water in the 5+ gallon reservoir after each run.

    d. Record drain time, discharged water weight, and remaining reservoir height after each run.

7. Repeat step six around 7–8 times.

C. <u>Intermediate Calculations</u>

1. Using measured values for spigot and siphon tube diameter, calculate spigot and siphon outlet areas using equation for area:

$$A = \frac{\pi}{4}D^2 \tag{8}$$

Result is to be used in Equation 6 to calculate theoretical volume flow rate.

2. Using measured values for discharged water weight and discharge time, calculate mass flow rate using equation:

$$\dot{m} = \frac{W_{discharged\ water}}{\Delta t_{discharged\ water}} \tag{9}$$

Result is to be used in Equation 7 to calculate experimental volume flow rate.

[This is very well done. I particularly like the use of the "C" section to describe how the data will be processed for use in calculating the results. 20/20 points]

RESULTS

This section includes resultant values both measured during experimentation and calculation results obtained using said measured values.

TABLE 1 **Measured Diameters and Calculated Area**

Item	Inner Diameter, D (m)	Calculated Exit Area (m^2)
Siphon Tube	0.008	$5.027*10^{-5}$ m
Spigot	0.00905	$6.433*10^{-5}$ m

[The diameter of the siphon tube and the spigot was taken with the same instrument and therefore should be reportable to the same number of significant figures. Not doing so introduces a disparity in the results of the two tests.] The inner diameters were measured using Vernier caliper in mm and converted into SI units (m). Equation 8 was used to calculate exit area.

TABLE 2 **Measured Quantities for Spigot Flow**

Run Number	Reservoir Water Height (in)	Discharge Time (s)	Receptacle Water Weight (kg)
1	12.25	17.12	0.658
2	11.25	31.20	1.602
3	10	25.82	1.112
4	8.875	32.56	1.430
5	7.5	27.99	1.010
6	6.375	34.43	1.234
7	5.25	35.66	1.114
8	4	40.90	1.098

The values in Table 2 were used in Equations 6, 7, and 9 to calculate mass flow rate, theoretical and experimental volume flow rate for discharge through the valve controlled spigot. Height values were converted into standard SI units (m). Receptacle water weight was calculated by subtracting the measured water weight by the dry mass of the receptacle bucket (1.008 kg). Run number is included to corresponding values from different tables.

TABLE 3 Measured Quantities for Siphon Flow

Run Number	Reservoir Water Height (in)	Discharge Time (s)	Receptacle Water Weight (kg)
1	12.25	36.49	1.782
2	10.75	30.35	1.258
3	9.5	40.11	1.808
4	8.25	30.23	1.000
5	7.25	31.91	1.012
6	6.125	37.81	0.910
7	5.125	42.51	1.230
8	2.75	40.25	0.812

The values in Table 2 were used in Equations 6, 7, and 9 to calculate mass flow rate, theoretical and experimental volume flow rate for discharge through the siphon tube. Height values were converted into standard SI units (m). Receptacle water weight was calculated by subtracting the measured water weight by the dry mass of the receptacle bucket (1.008 kg). Run number is included to corresponding values from different tables.

TABLE 4 Calculated Theoretical Volume Flow Rate for Spigot Flow

Run Number	Volume Flow Rate (m^3/s)
1	$1.242*10^{-4}$
2	$1.204*10^{-4}$
3	$1.122*10^{-4}$
4	$1.057*10^{-4}$
5	$9.718*10^{-5}$
6	$8.960*10^{-5}$
7	$8.131*10^{-5}$
8	$7.097*10^{-5}$

The theoretical volume flow rates were calculated using data from Tables 1, 2 and Equation 6.

TABLE 5 Calculated Theoretical Volume Flow Rate for Siphon Tube Flow

Run Number	Volume Flow Rate (m³/s)
1	$1.590*10^{-4}$
2	$1.489*10^{-4}$
3	$1.400*10^{-4}$
4	$1.304*10^{-4}$
5	$1.223*10^{-4}$
6	$1.124*10^{-4}$
7	$1.028*10^{-4}$
8	$8.794*10^{-5}$

The theoretical volume flow rates were calculated using data from Tables 1, 3 and Equation 6.

TABLE 6 Calculated Experimental Volume Flow Rate for Spigot Flow

Run Number	Discharged Water Weight (kg)	Time (s)	Mass Flow Rate (kg/s)	Volume Flow Rate (m³/s)
1	0.658	17.120	0.038435	$3.8435*10^{-5}$
2	1.602	31.200	0.051346	$5.1346*10^{-5}$
3	1.112	25.820	0.043067	$4.3067*10^{-5}$
4	1.430	32.560	0.043919	$4.3919*10^{-5}$
5	1.010	27.990	0.036084	$3.6084*10^{-5}$
6	1.234	34.425	0.035846	$3.5486*10^{-5}$
7	1.114	35.655	0.031244	$3.1244*10^{-5}$
8	1.098	40.897	0.026881	$2.6881*10^{-5}$

The experimental values for Mass Flow Rate and Volume Flow Rate were calculated using data from columns one and two (discharged water weight, time), and $\rho_{water} = 1000$ kg/m³. Equations 7 and 9 were used.

TABLE 7 Calculated Experimental Volume Flow Rate for Siphon Flow

Run Number	Discharged Water Weight (kg)	Time (s)	Mass Flow Rate (kg/s)	Volume Flow Rate (m³/s)
1	1.782	36.49	0.048835	$4.8835*10^{-5}$
2	1.258	30.35	0.041450	$4.1450*10^{-5}$
3	1.808	40.11	0.045076	$4.5076*10^{-5}$
4	1.000	30.23	0.033080	$3.3080*10^{-5}$
5	1.012	31.91	0.031714	$3.1714*10^{-5}$
6	0.910	37.81	0.024068	$2.4068*10^{-5}$
7	1.230	42.51	0.028934	$2.8934*10^{-5}$
8	0.812	40.25	0.020174	$2.0174*10^{-5}$

The experimental values for Mass Flow Rate and Volume Flow Rate were calculated using data from columns one and two (discharged water weight, time), and ρ_{water} = 1000 kg/m³. Equations 7 and 9 were used.

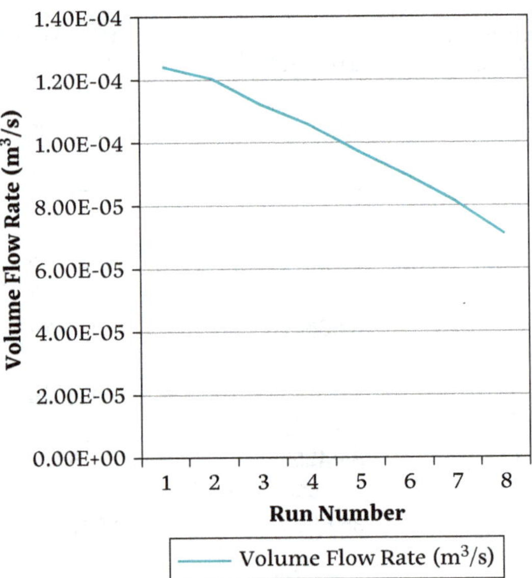

FIGURE 1 Theoretical Volume Flow Rate For Spigot Flow (m³/s)

[This plot of Vdot should be against height, not run number. Additionally, you haven't explained or justified the height that would be used in the plots or in the preceding tables.]

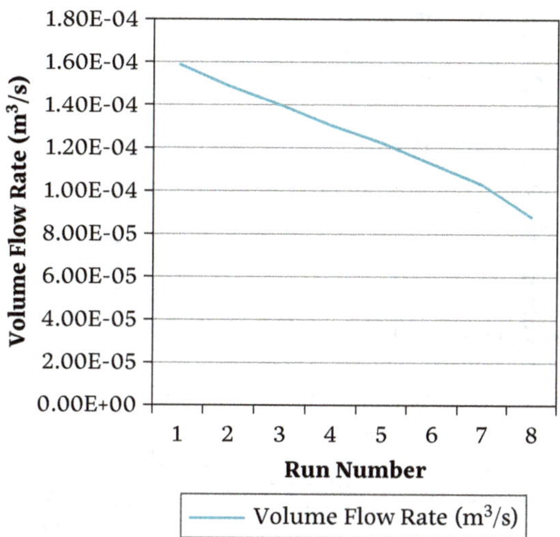

FIGURE 2 Theoretical Volume Flow Rate For Siphon Flow (m³/s)

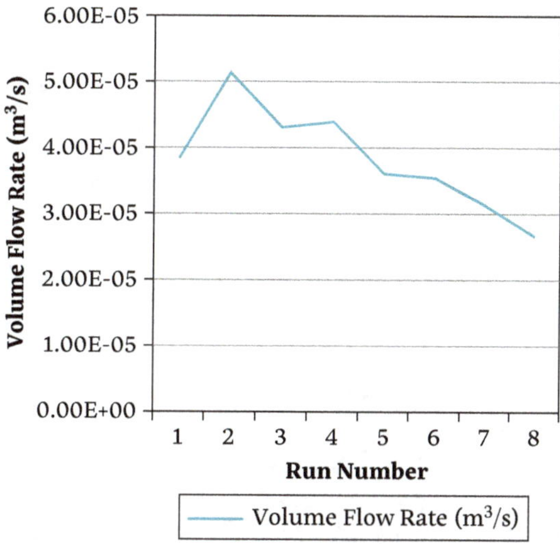

FIGURE 3 Experimental Volume Flow Rate For Spigot Flow (m³/s)

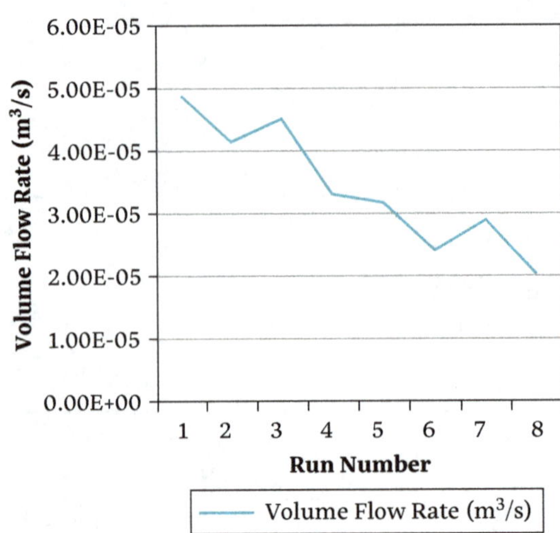

FIGURE 4 Experimental Volume Flow Rate for Siphon Flow (m³/s)

[The flow rates should never increase with a decrease in reservoir elevation. Measurements that suggests such illustrate a calculation or measurement error and should be explained.]

13/15 points

DISCUSSION AND CONCLUSION

The theoretical values calculated for Volume flow rate for discharge through both the siphon and spigot were on average both 300% higher than the values calculated through the use of experimental data. This is a significant difference in values. In addition, there was a noticeable difference in the values for discharge through the spigot and discharge through the siphon tube. The theoretical values were calculated as volume flow rate multiplied by exit area. The experimental values were calculated by simply dividing a known mass of water by a known time interval to obtain mass flow rate. From there, volume flow rate is calculated by simply dividing mass flow rate by the ~~viscosity~~ density (kg/m^3) of water. For the theoretical values for both spigot and siphon flow the area was calculated using the measured inner diameter of each. The theoretical value for exit velocity was calculated using Torricelli's Equation. Torricelli's Equation is derived from Bernoulli's Equation after some assumptions have been made. Torricelli's Equation states that Exit Velocity is equal to the square root of 2 times gravity times height.

The most likely reason for the big difference in experimental and theoretical values for volume flow rate is the use of Bernoulli's Equation and, after assumptions and manipulation, Torricelli's Equation. Bernoulli's Equation is limited in its use and requires the following conditions: steady flow, negligible viscous effects, no shaft work, incompressible flow, negligible heat transfer, and flow along a stream line [2]. Most of these conditions would be applicable in the flow rate experiment except for the negligible viscous effect. Every flow involves some friction, no matter how small, and frictional effects may or may not be negligible [3]. Bernoulli's Equation is a modified version of the Energy Equation. The Energy Equation adds terms for head loss due to friction, and head loss terms for turbines, pumps and other mechanical devices that produce and consume energy. I believe that the exclusion of the head loss term due to friction makes up the majority of the difference between the theoretical and experimental values for Volume Flow Rate. [True.] The remainder I believe is due to the use of Torricelli's Equation to calculate exit velocity. By cancelling out many of the Bernoulli terms, less accurate values result. [Agreed. There is some error associated with the assumption that V1 = 0.] A more accurate theoretical value for V2 would be obtained through the integration of the differential conservation of mass equation. Torricelli's equation was used instead for the following reasons: a diameter measurement for the reservoir bucket was not collected during experimentation, and the initial mass of the reservoir bucket with water and dry were also not obtained. While it would be possible to calculate these values using information gained through the lab, the process would have involved a few assumptions (such as the bucket initially held exactly five gallons) and complex calculations that would have introduced error into these values. These errors and the error from the use of the Torricelli equation would be similar, so the simpler equation was used. The difference in the values for spigot flow and siphon flow are most likely a result of the fact that the water flowing out of the siphon had to overcome some gravitational force to go up and over the bucket before exiting. I think this force caused the larger diameter siphon tube to have a smaller volume flow rate.

[I was about to state that a third source of error was the assumption of the height used in Torricelli's Equation. The midpoint would be a reasonable approximation. The beginning height overpredicts the exit velocity, the end height would be the lowest exit velocity. The integration would be the most accurate, as you have stated.]

[I think that the second and third errors you have identified would reduce the error to about 100% between theory and experiment.]

This experiment was successful in demonstrating the limitations of Bernoulli and Torricelli's equations when calculating velocity and the noticeable effect friction has on fluid flow.

[19/20 points]

REFERENCES

[1] Çengel, Y., Cimbala, J.,2014, *Fluid Mechanics: Fundamentals and Applications, Third Edition*. McGraw-Hill, New York, pp. 192–222, Chap. 5.
[2] De Nevers, N., & Grahn, R. (1991). *Fluid mechanics for chemical engineers*. McGraw-Hill, pp. 412–419.
[3] Van Driest, E. R. (2012). On turbulent flow near a wall. *Journal of the Aeronautical Sciences (Institute of the Aeronautical Sciences), 23*(11).

CPSIA information can be obtained
at www.ICGtesting.com
Printed in the USA
LVHW062112141021
700466LV00002B/3